GAME THEORY AND STRATEGY

by

Philip D. Straffin

Beloit College

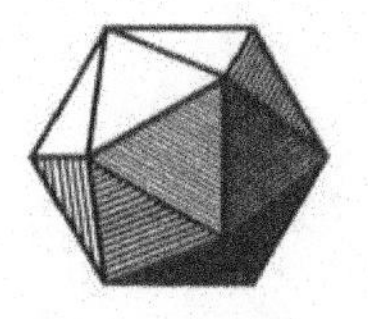

36

THE MATHEMATICAL ASSOCIATION OF AMERICA

Sixth printing: November 2017

Published in Washington, DC, by The Mathematical Association of America

Library of Congress Catalog Card Number 92-064176

Print ISBN 978-0-88385-637-6

Electronic ISBN 978-0-88385-950-6

Manufactured in the United States of America

NEW MATHEMATICAL LIBRARY

published by

The Mathematical Association of America

The New Mathematical Library (NML) was begun in 1961 by the School Mathematics Study Group to make available to high school students short expository books on various topics not usually covered in the high school syllabus. In three decades the NML has matured into a steadily growing series of some thirty titles of interest not only to the originally intended audience, but to college students and teachers at all levels. Previously published by Random House and L. W. Singer, the NML became a publication series of the Mathematical Association of America (MAA) in 1975. Under the auspices of the MAA the NML will continue to grow and will remain dedicated to its original and expanded purposes.

Note to the Reader

This book is one of a series written by professional mathematicians in order to make some important mathematical ideas interesting and understandable to a large audience of high school students and laymen. Most of the volumes in the *New Mathematical Library* cover topics not usually included in the high school curriculum; they vary in difficulty, and, even within a single book, some parts require a greater degree of concentration than others. Thus, while you need little technical knowledge to understand most of these books, you will have to make an intellectual effort.

If you have so far encountered mathematics only in classroom work, you should keep in mind that a book on mathematics cannot be read quickly. Nor must you expect to understand all parts of the book on first reading. You should feel free to skip complicated parts and return to them later; often an argument will be clarified by a subsequent remark. On the other hand, sections containing thoroughly familiar material may be read very quickly.

The best way to learn mathematics is to *do* mathematics, and each book includes problems some of which may require considerable thought. You are urged to acquire the habit of reading with paper and pencil in hand; in this way, mathematics will become increasingly meaningful to you.

The authors and editorial committee are interested in reactions to the books in this series and hope that you will write to: Anneli Lax, Editor, New Mathematical Library, New York University, The Courant Institute of Mathematical Sciences, 251 Mercer Street, New York, N.Y. 10012.

The Editors

NEW MATHEMATICAL LIBRARY

1. Numbers: Rational and Irrational *by Ivan Niven*
2. What is Calculus About? *by W. W. Sawyer*
3. An Introduction to Inequalities *by E. F. Beckenbach and R. Bellman*
4. Geometric Inequalities *by N. D. Kazarinoff*
5. The Contest Problem Book I Annual High School Mathematics Examinations 1950–1960. Compiled and with solutions *by Charles T. Salkind*
6. The Lore of Large Numbers *by P. J. Davis*
7. Uses of Infinity *by Leo Zippin*
8. Geometric Transformations I *by I. M. Yaglom, translated by A. Shields*
9. Continued Fractions *by Carl D. Olds*
10. Replaced by NML-34
11. } Hungarian Problem Books I and II, Based on the Eötvös Competitions
12. } 1894–1905 and 1906–1928, *translated by E. Rapaport*
13. Episodes from the Early History of Mathematics *by A. Aaboe*
14. Groups and Their Graphs *by E. Grossman and W. Magnus*
15. The Mathematics of Choice *by Ivan Niven*
16. From Pythagoras to Einstein *by K. O. Friedrichs*
17. The Contest Problem Book II Annual High School Mathematics Examinations 1961–1965. Compiled and with solutions *by Charles T. Salkind*
18. First Concepts of Topology *by W. G. Chinn and N. E. Steenrod*
19. Geometry Revisited *by H. S. M. Coxeter and S. L. Greitzer*
20. Invitation to Number Theory *by Oystein Ore*
21. Geometric Transformations II *by I. M. Yaglom, translated by A. Shields*
22. Elementary Cryptanalysis—A Mathematical Approach *by A. Sinkov*
23. Ingenuity in Mathematics *by Ross Honsberger*
24. Geometric Transformations III *by I. M. Yaglom, translated by A. Shenitzer*
25. The Contest Problem Book III Annual High School Mathematics Examinations 1966–1972. Compiled and with solutions *by C. T. Salkind and J. M. Earl*
26. Mathematical Methods in Science *by George Pólya*
27. International Mathematical Olympiads—1959–1977. Compiled and with solutions *by S. L. Greitzer*
28. The Mathematics of Games and Gambling *by Edward W. Packel*
29. The Contest Problem Book IV Annual High School Mathematics Examinations 1973–1982. Compiled and with solutions *by R. A. Artino, A. M. Gaglione, and N. Shell*
30. The Role of Mathematics in Science *by M. M. Schiffer and L. Bowden*
31. International Mathematical Olympiads 1978–1985 and forty supplementary problems. Compiled and with solutions *by Murray S. Klamkin*
32. Riddles of the Sphinx *by Martin Gardner*
33. U.S.A. Mathematical Olympiads 1972–1986. Compiled and with solutions *by Murray S. Klamkin*
34. Graphs and Their Uses *by Oystein Ore.* Revised and updated *by Robin J. Wilson*
35. Exploring Mathematics with Your Computer *by Arthur Engel*
36. Game Theory and Strategy *by Philip D. Straffin, Jr.*

37. Episodes in Nineteenth and Twentieth Century Euclidean Geometry *by Ross Honsberger*
38. The Contest Problem Book V American High School Mathematics Examinations and American Invitational Mathematics Examinations 1983–1988. Compiled and augmented *by George Berzsenyi and Stephen B. Maurer*
39. Over and Over Again *by Gengzhe Chang and Thomas W. Sederberg*
40. The Contest Problem Book VI American High School Mathematics Examinations 1989–1994. Compiled and augmented *by Leo J. Schneider*
41. The Geometry of Numbers *by C.D. Olds, Anneli Lax, and Giuliana Davidoff*
42. Hungarian Problem Book III Based on the Eötvös Competitions 1929–1943 *translated by Andy Liu*
43. Mathematical Miniatures *by Svetoslav Savchev and Titu Andreescu*

Other titles in preparation.

Books may be ordered from:
MAA Service Center
P. O. Box 91112
Washington, DC 20090-1112
1-800-331-1622 fax: 301-206-9789

Contents

Preface

Given the long history of mathematics, and the longer history of human conflict, the mathematical theory of conflict known as game theory is a surprisingly recent creation. John von Neumann published the fundamental theorem of two-person zero-sum games in 1928, but at that time it was simply a theorem in pure mathematics. The theory which provided a broader intellectual context, game theory, was begun in an intense period of collaboration between von Neumann and the economist Oskar Morgenstern in the early 1940's. Their work appeared as the monumental *Theory of Games and Economic Behavior* [1944].

The importance of the new subject and its potential for social science were quickly recognized. A. H. Copeland [1945] wrote:

> Posterity may regard this book as one of the major scientific achievements of the first half of the twentieth century. This will undoubtedly be the case if the authors have succeeded in establishing a new exact science—the science of economics. The foundation which they have laid is extremely promising.

In the years since 1944 game theory has developed rapidly. It now has a central place in economic theory, and it has contributed important insights to all areas of social science.

For a number of years I have taught a course titled "Game Theory and Strategy" in Beloit College's interdisciplinary division. About half of the students in the course have been mathematics and science majors; the other half were majors in areas like economics, government, and psychology. My aim has been to teach the most important ideas of mathematical game theory in an interdisciplinary context. This book grew out of that course.

The interdisciplinary context of the course requires that mathematical prerequisites be kept to a minimum. High school mathematics through algebra and elementary analytic geometry (the equation of a line in the coordinate plane) will suffice. On the other hand, I think you will find the ideas of game theory interesting and challenging. You should be prepared to think carefully and precisely.

The interdisciplinary context also requires that serious attention be paid to the uses of game-theoretic ideas in other disciplines. In this book, chapters presenting mathematical theory are interspersed with chapters presenting applications of that theory in anthropology, social psychology, economics, politics, business, biology and philosophy. I believe that the breadth and depth of applications here is greater than in any other elementary treatment of game theory, and that the applications give an idea of the impact which game-theoretic thinking has had on modern social thought.

Finally, the course "Game Theory and Strategy" is an active course. Students don't just hear about game theory and its uses; they play matrix games and coalition games, discuss ideas, work many problems, and write about applications. I think that you will get the most out of this book if you approach it in a similarly active way. Look up some of the references. Talk about these ideas with your friends; try them out on your enemies. Above all, do the exercises at the end of the chapters. I have tried to make them both illuminating and enjoyable, and there are answers in the back of the book.

Many people have contributed to the development of my interest in game theory. William Lucas and his colleagues in the Department of Operations Research at Cornell University provided a stimulating sabbatical environment when I was first beginning to think about these ideas. I have learned from collaboration on game-theoretic applications with Steven Brams, Morton Davis, James Heaney, and Bernie Grofman. The Rockefeller Foundation funded a research term in the Department of Environmental Engineering at the University of Florida. This book was begun during a sabbatical at the Kennedy School of Government and the Harvard Business School, where I had the pleasure of talking with Howard Raiffa and his colleagues. I am grateful to all of them.

However, the people who have contributed most to the organization and presentation of the ideas in this book are the students in my game theory class over the past fifteen years. I have benefited enormously from their patience, criticisms, ideas, and enthusiasm. This book is dedicated to them.

Philip Straffin, 1993

Part I

Two-Person Zero-Sum Games

1. The Nature of Games

Game theory is the logical analysis of situations of conflict and cooperation. More specifically, a *game* is defined to be any situation in which

i) There are at least two *players*. A player may be an individual, but it may also be a more general entity like a company, a nation, or even a biological species.
ii) Each player has a number of possible *strategies*, courses of action which he or she may choose to follow.
iii) The strategies chosen by each player determine the *outcome* of the game.
iv) Associated to each possible outcome of the game is a collection of numerical *payoffs*, one to each player. These payoffs represent the value of the outcome to the different players.

In the course of our study, all of these components of a game—players, strategies, outcomes, and payoffs—will acquire additional richness and precision. For the moment we will be content with an intuitive understanding of these ideas.

Game theory is the study of how players should rationally play games. Each player would like the game to end in an outcome which gives him as large a payoff as possible. He has some control over the outcome, since his choice of strategy will influence it. However, the outcome is not determined by his choice alone, but also depends upon the choices of all the other players, and this is where the conflict and cooperation enter. There may be conflict because different players will, in general, value outcomes differently. There is a chance for cooperation because several players together may be able to coordinate their strategies to obtain an outcome with better payoffs for all of them. Rational play will involve complicated individual decisions about how to choose a strategy which will produce an outcome favorable to you, knowing that other players are trying to choose strategies which will produce an outcome favorable to them. It will also involve social decisions about how and with whom to try to cooperate.

Traditional games like chess, bridge and poker are certainly games in our sense. Game theory is so named, of course, because it abstracts from and generalizes the study of these kinds of traditional games. The abstraction and generalization is powerful enough to include a wide variety of important social situations. Companies pursuing "corporate strategies" are playing a game. So are political candidates trying to win an election, members of Congress trying to pass or defeat a bill, nations maneuvering in the international arena. We will see that members of a biological species can be thought of as players in a game in which the payoff is the chance to pass along genes to future generations. If we can develop a general theory of how to play games rationally, it should have important and widespread consequences.

This last statement might conjure up grand visions of a theory which will prescribe the best course of action in any situation of conflict and cooperation. I should say at the outset that the goal of game theory is considerably more modest. At least three serious obstacles preclude the development of any unified prescriptive theory. First, any real-world game is enormously complex. It may be hard to say who the players are, it is usually impossible to delineate all conceivable strategies and say what outcomes they lead to, and it is not easy to assign payoffs to any given outcome. The best we can hope to do is to build a simple game which *models* some important features of the real situation. The building of this model, and its analysis, may give insight into the original situation.

The second obstacle is that game theory deals with play which is *rational*. Each player logically analyzes the best way to achieve her ends, given that the other players are logically analyzing the best way to achieve their ends. In other words, rational play assumes rational opponents. In the real world, it is quite doubtful that all players will play rationally. However, the extent to which players do or do not behave rationally in situations of conflict and cooperation is an interesting question in itself, and one on which game theory can shed light.

Perhaps the most serious obstacle is that, as we will see, game theory does not have a unique prescription for play in games with two players whose interests are not completely opposed. It also does not have such a prescription for games with more than two players. What game theory offers is a variety of interesting examples, analyses, suggestions and partial prescriptions for these situations.

To get an idea of the kinds of situations we will be dealing with, let's look at some simple examples. In Parts I and II of this book, we will be concerned with games in which there are just two players. Call them "Rose" and "Colin."[†] Suppose Rose has three strategies, and Colin has two, so there are six different possible outcomes. We can represent this situation by a 3×2 array. We can think of Rose choosing a strategy, i.e., one of the three rows, and writing it on a piece of paper which she places on the table, face down. Colin picks a column in the

		Colin A	Colin B
Rose	A	$(2, -2)$	$(-3, 3)$
	B	$(0, 0)$	$(2, -2)$
	C	$(-5, 5)$	$(10, -10)$

Game 1.1

[†]I am indebted to Peter Ungar for introducing me to Rose, who chooses among the rows of a game matrix, and Colin, who chooses among the columns.

same way. Then they turn over their papers and determine their payoffs. For each outcome, the first number is the payoff to Rose, the second the payoff to Colin.

For this game, notice that the two payoffs for each outcome add to zero. This means that whatever Rose gains Colin loses, and visa versa, so the interests of Rose and Colin are strictly opposed. Such a game is called a *zero-sum game.* It represents a situation of pure conflict between our two players. For a two-person zero-sum game, it is enough to give the payoff to Rose for each outcome, since the payoff to Colin will just be the corresponding negative. Hence we can represent the game as shown:

		Colin	
		A	B
Rose	A	2	−3
	B	0	2
	C	−5	10

Game 1.1 represented by payoffs to Rose

Rose wants an outcome with a high number; Colin wants an outcome with a low number. How should they play?

Suppose Rose reasons that she would most like 10, so she should play Rose C, hoping that Colin will play Colin B. The problem is that if Colin knows or guesses that Rose will do this, Colin will play Colin A and Rose will get −5. On the other hand, if Rose foresees this she can play Rose A and get 2. Foreseeing this, Colin should play Colin B, but then Rose should play Rose C... Round and round we go. We can diagram the motion like this:

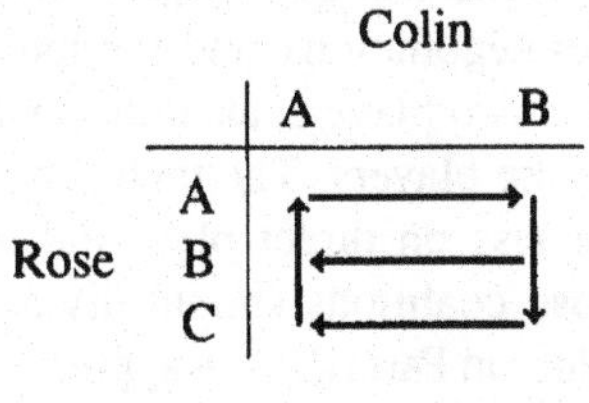

		Colin	
		A	B
Rose	A	↑ →	→ ↓
	B	↑ ←	← ↓
	C	←	← ↓

Movement diagram for Game 1.1

Without reference to guessing, the arrows are drawn as follows. In each row we draw an arrow from each entry to the smallest entry in that row. In each column we draw arrows to the largest entry in that column.

In Game 1.1, there is no stopping point, and play seems to hinge on how far the players can go in a chain of reasoning like, "If I think that he thinks that I think that he will do that, then I should do this." In Part I of this book we will see how to resolve this dilemma by using the idea of "mixed strategies" to completely prescribe rational play in any two-person zero-sum game. We will also look more closely at the notion of a strategy by studying *game trees*, and at the idea of a payoff, studying *utility theory*. We will consider applications to fishing strategies in Jamaica, warfare, business, and the philosophical problem of free will.

The next most complicated kind of situation is where there are still just two players, but the payoffs for outcomes do not always add to zero. For example,

		Colin A	Colin B
Rose	A	$(1, 1)$	$(-2, 2)$
	B	$(2, -2)$	$(-5, -5)$

Game 1.2

What should the players do here? The most attractive outcome seems to be $(1, 1)$, which is second best for both players and has the highest total payoff. But if Colin knows that Rose will choose Rose A, Colin will do better by playing Colin B. Similar reasoning holds for Rose. If both players reason like this, they will end up at $(-5, -5)$, the worst outcome for both of them.

Of course, if the players can communicate, they can agree to both choose A. Or can they? Colin announces, "I am going to choose Colin B," and hangs up the telephone. You are Rose. What do you do? Perhaps you wish you had thought of doing that first.

In Part II, we will consider two-person non-zero-sum games, both without communication and with communication, and from the point of view of an arbitrator trying to recommend a fair settlement of such a game. We shall not find a general solution for such games, but we will discover a number of interesting things about them. Applications will be to experimental social psychology, the evolution of behavior, labor negotiations, and economic duopolies.

When there are more than two players, an additional complication arises from possible coalitions among the players. The theory of N-person games ($N > 2$) has tended to concentrate less on direct play, and more on which coalitions should form and how those coalitions should divide their winnings. We will discuss N-person game theory in Part III, with applications to politics, economics, anthropology and athletics.

2. Matrix Games: Dominance and Saddle Points

We saw in Chapter 1 that a two-person zero-sum game where Rose has m strategies and Colin has n strategies can be represented by an $m \times n$ array of numbers, giving the payoffs from Colin to Rose for each of the $m \cdot n$ possible outcomes. Such an array is called an $m \times n$ matrix, so these games are also known as *matrix games*. Rose wishes to choose a row of the matrix which will result in a large number; Colin wishes to choose a column which will result in a small number.

Before reading further, I would suggest that you find a friend and try playing the following matrix game, say 20 times, recording the strategy choices and the payoffs to Rose:

		Colin			
		A	B	C	D
Rose	A	12	−1	1	0
	B	5	1	7	−20
	C	3	2	4	3
	D	−16	0	0	16

Game 2.1

Remember that Rose wants the payoffs to be large (16 would be best), while Colin wants the payoffs to be small (−20 would be best).

Well, how did it go? In my most recent Game Theory class, the results for plays 11–20 were like this:

Rose Strategy	Percent of time	Colin Strategy	Percent of time
A	31%	A	20%
B	10%	B	51%
C	49%	C	2%
D	10%	D	27%
	100%		100%

The average total payoff to Rose, over 20 games, was 39. The game seems to be biased in Rose's favor.

The first thing to notice about these results is the very low incidence of Colin C. If you played Colin in the experiment, I expect you can explain why you would not use Colin C, at least after you got used to the game. One reason might be that there aren't any negative numbers under Colin C. A better reason is that

Colin B is *strictly better* than Colin C, since all of the numbers in the second column are less than or equal to the corresponding numbers in the third column. We say that Colin B *dominates* Colin C, or that Colin C is *dominated by* Colin B. Clearly it never makes sense to play a strategy which is dominated by some other strategy: you would always do as well or better by playing the dominating strategy. In general,

DEFINITION. A strategy S *dominates* a strategy T if every outcome in S is at least as good as the corresponding outcome in T, and at least one outcome in S is strictly better than the corresponding outcome in T.

DOMINANCE PRINCIPLE. A rational player should never play a dominated strategy.

The Dominance Principle may eliminate some strategies, but its usefulness is often limited. In Game 2.1, you can check that none of Colin A, B and D dominates any of the others, and none of Rose's strategies dominates any other Rose strategy. Hence we cannot use the principle to restrict strategy choices further.

The second thing to notice about the experimental results is that Rose C and Colin B were played far more than any of the other strategies. Why is this? One reason might be that these are the *most cautious* strategies for Rose and Colin. With Rose C, Rose is assured of winning at least 2, whereas with any other strategy she may in fact lose. With Colin B, Colin is assured of losing no more than 2, whereas with any other strategy his loss may be larger.

However, there are good reasons for the players to choose Rose C–Colin B even if they do not feel particularly cautious. Here are two of them:

1) Rose C–Colin B is an *equilibrium outcome*. This means that if Colin knows or believes that Rose will play Rose C, Colin would want to respond with Colin B, and similarly, Rose C is Rose's best response to Colin B. If both players are playing these strategies, neither player has any incentive to move to a different strategy. This can be nicely seen from the movement diagram of the game:

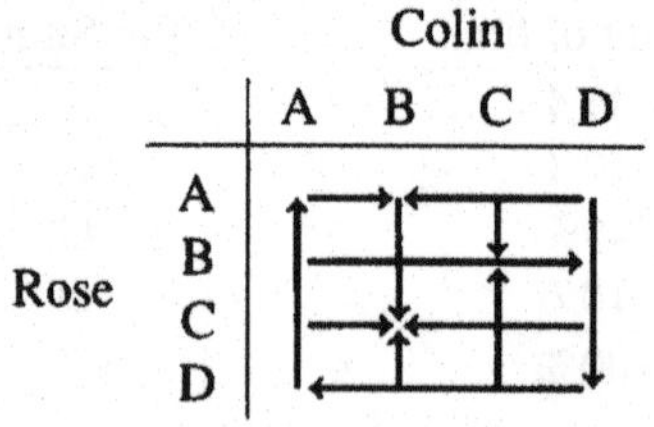

Colin
A B C D
Rose A B C D

2) By playing Rose C, Rose can assure that she will win *at least two* units. By playing Colin B, Colin can assure that Rose will win *no more than two* units. If Rose wins less than 2, she could have done better by playing Rose C; if Rose wins more than 2, Colin could have done better by playing Colin B. If the players are not playing their equilibrium strategies, one or the other knows that he or she could have forced a better outcome.

Game theorists consider these two arguments powerful enough to prescribe Rose C–Colin B as rational play in this game. The arguments follow from the fact that the payoff at Rose C–Colin B is simultaneously the smallest number in its row and the largest number in its column. This means that the strategies leading to this outcome are best responses to each other, and that both players are guaranteed to do no worse than this value. In general, we have

DEFINITION. An outcome in a matrix game (with payoffs to the row player) is called a *saddle point*† if the entry at that outcome is both less than or equal to any entry in its row, and greater than or equal to any entry in its column.

SADDLE POINT PRINCIPLE. If a matrix game has a saddle point, both players should play a strategy which contains it.

DEFINITION. For any matrix game, if there is a number v such that Rose has a strategy which guarantees that she will win at least v, and Colin has a strategy which guarantees that Rose will win no more than v, then v is called the *value* of the game.

If a game has a saddle point, the saddle point entry is the value of the game. We will see below that a matrix game may not have a saddle point. It is also possible that a game could have more than one saddle point:

		Colin			
		A	B	C	D
Rose	A	4	(2)	5	(2)
	B	2	1	−1	−20
	C	3	(2)	4	(2)
	D	−16	0	16	1

Game 2.2

All four of the circled outcomes are saddle points. Notice that they are the only saddle points. In particular, the "2" at Rose B–Colin A is not a saddle point.

† The term "saddle point" comes from a picture of the condition that the entry is smallest in its row and largest in its column. If we drew a three-dimensional graph of the payoffs, the part around a saddle point would be shaped like a saddle:

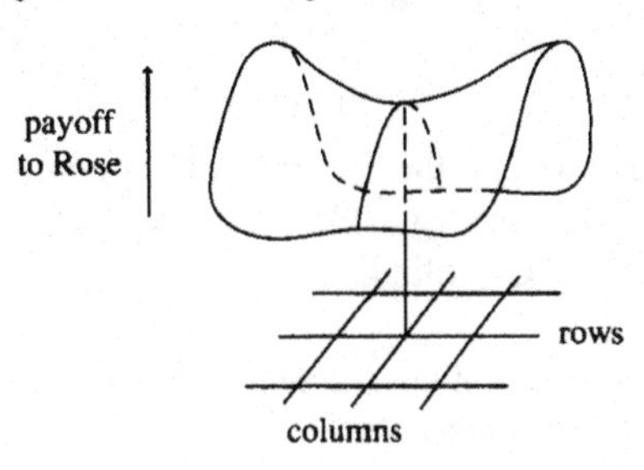

These multiple saddle points are very nicely related: they all have the same value, and they appear at the corners of a rectangle. This relation holds in any matrix game which has multiple saddle points:

THEOREM. Any two saddle points in a matrix game have the same value. Furthermore, if Rose and Colin both play strategies containing a saddle point outcome, the result will always be a saddle point.[†]

PROOF. Suppose that a and b are saddle point entries in a matrix game, and c and d are the other entries at the corners of a rectangle containing a and b:

$$\begin{matrix} a & \cdots & c \\ \vdots & & \vdots \\ d & \cdots & b \end{matrix}$$

Since a is a smallest entry in its row and b is a largest entry in its column, we get $a \leqslant c \leqslant b$. Since b is a smallest entry in its row and a is a largest entry in its column, we get $b \leqslant d \leqslant a$. Putting these together, we see that all the inequalities must in fact be equalities, so that all four numbers are the same. Hence c and d are also largest in their columns and smallest in their rows, and thus are saddle points. Q.E.D.

We need a way of telling whether a game has a saddle point, and find the saddle point (or points) if it does. Of course, we could just look at every entry in the matrix, checking if it is smallest in its row and largest in its column. However, there is a more efficient way, based on the idea that saddle point strategies, if they exist, will be "most cautious" strategies. The method, illustrated below, is first to write down the minimum entry in each row, and circle the maximum of these row minima. Then write down the maximum entry in each column, and circle the minimum of these column maxima.

		Colin					
		A	B	C	D	Row minimum	
Rose	A	4	3	(2)	5	2	← maximin
	B	−10	2	0	−1	−10	
	C	7	5	(2)	3	2	← maximin
	D	0	8	−4	−5	−5	
Column maximum		7	8	2	5		
				↑ minimax			

If the maximin of the rows and the minimax of the columns are the *same*, then they appear at saddle point strategies. In the example, the two saddle points are Rose A–Colin C and Rose C–Colin C.

[†] We will refer to the two results of this theorem as saying that saddle points in a matrix game are *equivalent* and *interchangeable*.

In some games the row maximin and the column minimax are not the same. For example, consider the game from Chapter 1:

		Colin A	Colin B	Row minimum	
Rose	A	2	−3	−3	
Rose	B	0	2	0	← maximin
Rose	C	−5	10	−5	
Column maximum		2	10		
		↑ minimax			

Game 1.1

When the maximin and the minimax are different, the game has no saddle point. In Game 1.1, Rose can assure that she wins at least 0, and Colin can assure that Rose wins no more than 2, but the ground between 0 and 2 seems open for contest. Notice that the intersection of Rose's maximin strategy Rose B with Colin's minimax strategy Colin A is not an equilibrium outcome. In fact, we saw in Chapter 1 that this game has no equilibrium outcome. We will find a different way to deal with it in the next chapter.

Exercises for Chapter 2

1. In the following game, find all cases of dominance among Rose strategies and among Colin strategies.

		Colin A	Colin B	Colin C	Colin D
Rose	A	3	−6	2	−4
Rose	B	2	1	0	1
Rose	C	−4	3	−5	4

2. The Dominance Principle can be extended to the Principle of Higher Order Dominance. The idea is that we first cross out any dominated strategies for Rose and for Colin. In the resulting smaller game, some strategies may be dominated, even though they weren't in the original game. Cross them out. Look at the new smaller game and continue until no new dominance appears. The Principle of Higher Order Dominance says that players should play only strategies which survive this multi-stage process. Sometimes this principle can simplify a game enormously. In the following game, which strategies are admissible by the Principle of Higher Order Dominance?

		Colin A	Colin B	Colin C	Colin D	Colin E
Rose	A	1	1	1	2	2
Rose	B	2	1	1	1	2
Rose	C	2	2	1	1	1
Rose	D	2	2	2	1	0

3. Find all saddle points in the following games. Draw the movement diagrams for the games in b) and c).

a)

		Colin A	B	C	D
	A	3	2	4	2
Rose	B	2	1	3	0
	C	2	2	2	2

b)

	Colin A	B	C
A	−2	0	4
B	2	1	3
C	3	−1	−2

c)

	Colin A	B	C
A	4	3	8
B	9	5	1
C	2	7	6

4. The game in Exercise 3a shows that saddle points can appear in dominated strategies. The Dominance Principle says we shouldn't play these strategies. Show that the Dominance Principle cannot come into direct conflict with the Saddle Point Principle by showing that if Rose A dominates Rose B, and Rose B contains a saddle point entry b, then the entry a in the same column of Rose A is also a saddle point.

5. The efficient method for finding saddle points depends on two assertions I did not prove. Provide arguments for the following:
   a) If a is a saddle point entry, then the row containing a is a maximin row, the column containing a is a minimax column, and

$$\text{maximin} = a = \text{minimax}.$$

   b) If maximin = minimax, then the intersection of the maximin row and the minimax column is a saddle point.

6. a) In Game 2.1, suppose you were Rose and you knew that Colin was playing his strategies in the proportions given in my class results, and that what you did wouldn't change this. Which strategy would be best for you to play? Why? How would you calculate this?
   b) Same question if you were Colin and you knew Rose's proportions.

7. In the movement diagram for Game 2.1, notice that all movement tends to the saddle point: no matter where you start, if you follow the arrows you will eventually end up at the saddle point. Is this true for any game which has a saddle point?

3. Matrix Games: Mixed Strategies

We saw in Chapter 2 that in some matrix games the row maximin and the column minimax are different numbers, and in those games there is no saddle point. For example, consider

		Colin A	Colin B	Row minimum	
Rose	A	2	−3	−3	
	B	0	3	0	← maximin
Column maximum		2	3		
		↑ minimax			

Game 3.1

Since there is no saddle point in this game, neither player would want to play a single strategy with certainty, for the other player could take advantage of such a choice. The only sensible plan is to use some random device to decide which strategy to play. For example, Colin might flip a coin to decide between Colin A and Colin B. Such a plan, which involves playing a mixture of strategies according to certain fixed probabilities, is called a *mixed strategy*. The contrasting plan of playing one strategy with certainty is called a *pure strategy*.

To analyze the effect of one or both players using mixed strategies, we can use the concept of *expected value*.

DEFINITION. The *expected value* of getting payoffs $a_1, a_2, \ldots, a_k$ with respective probabilities $p_1, p_2, \ldots, p_k$ is $p_1a_1 + p_2a_2 + \cdots + p_ka_k$.

The expected value of a set of payoffs is just the weighted average of those payoffs, where the weights are the probabilities that each will occur. If Colin uses the coin-flipping mixed strategy in Game 3.1, he will play Colin A with probability $\frac{1}{2}$, Colin B with probability $\frac{1}{2}$. Hence if Rose plays Rose A, she will get a payoff of 2 with probability $\frac{1}{2}$, and a payoff of -3 with probability of $\frac{1}{2}$. Thus her expected payoff for Rose A is $\frac{1}{2}(2) + \frac{1}{2}(-3) = -\frac{1}{2}$. On the other hand, her expected payoff for Rose B is $\frac{1}{2}(0) + \frac{1}{2}(3) = \frac{3}{2}$. Clearly, if Rose knows or guesses that Colin is playing the mixed strategy $\frac{1}{2}$A, $\frac{1}{2}$B, Rose should play Rose B. This reasoning is known as the

EXPECTED VALUE PRINCIPLE. If you know that your opponent is playing a given mixed strategy, and will continue to play it regardless of what you do, you should play your strategy which has the largest expected value.

Now consider the situation from Colin's point of view. If Colin uses the mixed strategy $\frac{1}{2}$A, $\frac{1}{2}$B and Rose guesses this, Rose can take advantage of her knowledge to get an expected payoff of $\frac{3}{2}$. Colin might consider using mixed strategies with different probabilities, for example $\frac{1}{3}$A, $\frac{2}{3}$B. Might there be some choice of probabilities which Rose could not take advantage of? To find out, suppose that Colin plays a mixed strategy with probabilities x for A, $(1 - x)$ for B, where x is some number between 0 and 1. Calculate Rose's expected values for Rose A and Rose B:

Rose A: $x(2) + (1 - x)(-3) = -3 + 5x$

Rose B: $x(0) + (1 - x)(\ 3) = \ 3 - 3x$

Rose will not be able to take advantage of Colin's mixed strategy if these two expected values are the same: $-3 + 5x = 3 - 3x$. Solving, we get $x = \frac{3}{4}$. If Colin plays the mixed strategy $\frac{3}{4}$A, $\frac{1}{4}$B, Colin can assure that Rose wins, on average, no more than $\frac{3}{4}$ unit per game, regardless of how Rose plays:

Rose A: $\frac{3}{4}(2) + \frac{1}{4}(-3) = \frac{3}{4}$

Rose B: $\frac{3}{4}(0) + \frac{1}{4}(\ 3) = \frac{3}{4}$

It is worthwhile thinking a little about how Colin might in practice play a mixed strategy with these probabilities. Here are a few random suggestions:

- flip two coins; play B if they both come up tails.
- draw a card from a shuffled deck; play B if it is a spade.
- put three slips of paper marked A and one marked B in a hat and draw.
- look at the seconds reading on a digital watch; play B if it is divisible by four.
- have a computer generate a random number between 0 and 1; play B if it is larger than .75.

If you need to play a mixed strategy with unequal probabilities, there is plenty of room to use your ingenuity.

Now let's consider Game 3.1 from Rose's point of view. Trying to find a mixed strategy xA, $(1 - x)$B which Colin cannot take advantage of, Rose does the same kind of calculation:

Colin A: $x(2) + (1 - x)(0) = 2x$

Colin B: $x(-3) + (1 - x)(3) = 3 - 6x$

Setting $2x = 3 - 6x$, we solve for $x = \frac{3}{8}$. If Rose plays the mixed strategy $\frac{3}{8}$A, $\frac{5}{8}$B, she is assured of winning, on average, at least $\frac{3}{4}$ units per game, regardless of how Colin plays:

$$\text{Colin A:} \quad \frac{3}{8}(\ 2) + \frac{5}{8}(0) = \frac{3}{4}$$

$$\text{Colin B:} \quad \frac{3}{8}(-3) + \frac{5}{8}(3) = \frac{3}{4}$$

Comparing this with our earlier statement about Colin, we notice a strong similarity to the case of a game with a saddle point. Rose has a (mixed) strategy which ensures Rose an (expected) payoff of at least $\frac{3}{4}$; Colin has a (mixed) strategy which ensures that Rose's (expected) payoff will be no more than $\frac{3}{4}$. Reasoning as in the saddle point case, game theorists prescribe

- $\frac{3}{4}$ as the value of the game
- $\frac{3}{4}$A, $\frac{1}{4}$B as Colin's optimal strategy
- $\frac{3}{8}$A, $\frac{5}{8}$B as Rose's optimal strategy.

This value and these two optimal strategies are called the *solution* of the game. We will soon state a theorem which says that every matrix game has such a solution, either in pure strategies (a saddle point) or in mixed strategies.

There is a useful shorthand way to calculate mixed strategy solutions to 2×2 games without saddle points, which was popularized in [Williams, 1986] and is illustrated in the following diagram. To find the "oddments" with which Rose should play Rose A and Rose B, take the *absolute values* of the differences of the row entries, and *interchange* them. To find Colin's oddments, do the same with the differences of the column entries. You are asked in Exercise 1 to show that this method does indeed produce the correct probabilities.

		Colin A	Colin B	Row differences	Rose oddments	Rose probabilities
Rose	A	2	−3	2 − (−3) = 5	3	3/8
	B	0	3	0 − 3 = −3	5	5/8
Column differences		2 − 0 = 2	−3 − 3 = −6			
Colin oddments		6	2			
Colin probabilities		6/8	2/8			

It is important that you check for a saddle point before you use this method to find an optimal mixed strategy. Exercise 2 asks you to check that if a game has a saddle point, this method will not produce optimal strategies.

Beyond 2×2 games, the next most complicated games are $2 \times n$, where Rose has 2 pure strategies and Colin has $n > 2$, or $m \times 2$, where Rose has $m > 2$ pure strategies and Colin has 2. If such a game does not have a saddle point, it turns out that it always has a solution which is the mixed strategy solution to one of its 2×2 subgames. For large m or n, there could be a large number of 2×2 subgames to try. Fortunately, there is an elegant graphical technique to find which 2×2 subgame gives the solution to the game, while also giving a geometrical insight into what the solution means. I'll illustrate with two examples.

First, consider the 3×2 game from Chapter 1:

		Colin A	Colin B
Rose	A	2	−3
	B	0	2
	C	−5	10

Game 1.1

We have already checked that this game does not have a saddle point. Draw the graph in Figure 3.1a. For each Rose strategy, mark the payoff if Colin plays Colin A on the left axis, the payoff if Colin plays Colin B on the right axis, and connect them with a line. Notice that the vertical coordinate of this line above any point x gives Rose's expected payoff if Colin plays the mixed strategy $(1-x)\mathrm{A}, x\mathrm{B}$.

If Rose knew or guessed Colin's mixed strategy, she could take advantage by choosing her best response, which would mean that the outcome would lie on the *upper envelope* (the heavy line segments) of the graph. Colin would want

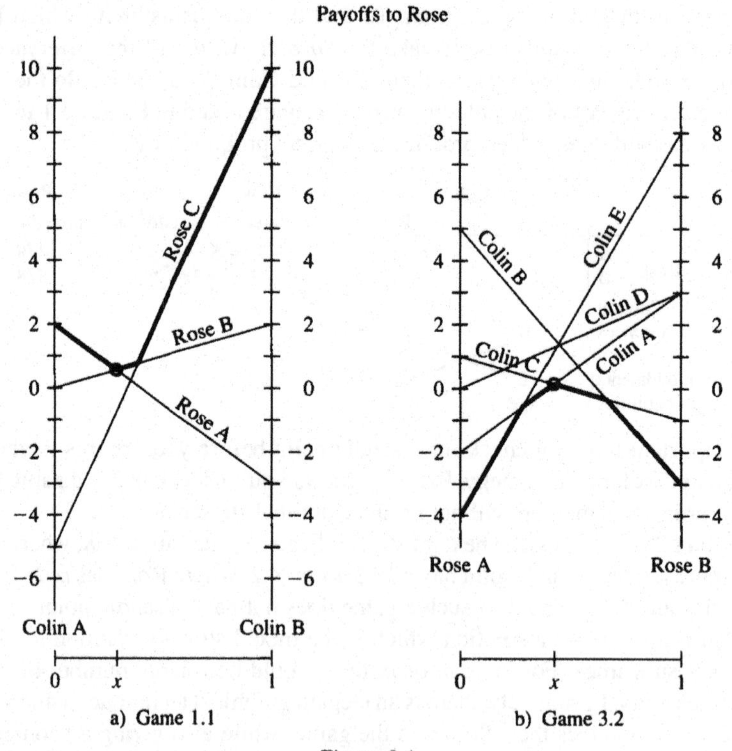

Figure 3.1

to choose x to make the corresponding payoff to Rose as small as possible—to get the *lowest point on the upper envelope*. Since this point (circled) is at the intersection of the lines for Rose A and Rose B, the appropriate subgame to solve is

		Colin A	Colin B	Rose probabilities
Rose	A	2	−3	2/7
	B	0	2	5/7
Colin probabilities		5/7	2/7	

Rose should play A with probability $\frac{2}{7}$, B with probability $\frac{5}{7}$, and should never play C. We can check that this is a solution of the game by doing the following calculation:

Rose expectations if Colin plays (5/7)A, (2/7)B

Rose A: $\frac{5}{7}(2) + \frac{2}{7}(-3) = \frac{4}{7}$

Rose B: $\frac{5}{7}(0) + \frac{2}{7}(2) = \frac{4}{7}$

Rose C: $\frac{5}{7}(-5) + \frac{2}{7}(10) = -\frac{5}{7}$

Colin expectations if Rose plays (2/7)A, (5/7)B, (0)C

Colin A: $\frac{2}{7}(2) + \frac{5}{7}(0) + (0)(-5) = \frac{4}{7}$

Colin B: $\frac{2}{7}(-3) + \frac{5}{7}(2) + (0)(10) = \frac{4}{7}$

The key result is that all of Rose's expectations are $\leqslant \frac{4}{7}$, and all of Colin's expectations are $\geqslant \frac{4}{7}$. Thus Colin assures that Rose will not win more than $\frac{4}{7}$, and Rose assures that Rose will not win less than $\frac{4}{7}$: the value of the game is $\frac{4}{7}$. Notice the interpretation of these results on the graph of Figure 3.1a. The lowest point on the upper envelope lies above $x = \frac{2}{7}$ ($\frac{5}{7}$ of the way toward Colin A). The vertical coordinate of this point is $\frac{4}{7}$, the value of the game. Above $x = \frac{2}{7}$, the line for Rose C has vertical coordinate $-\frac{5}{7}$, lower than the value of the game, which is why Rose C should not be used.

For $2 \times n$ games the same graphical technique works with one important change. Here is a quick example:

		Colin A	Colin B	Colin C	Colin D	Colin E
Rose	A	−2	5	1	0	−4
	B	3	−3	−1	3	8

Game 3.2

The graph for this game is shown in Figure 3.1b. Colin would try to choose his strategy to stay on the lower envelope, and Rose would want to choose x to get the *highest point on the lower envelope*. This involves Colin A and Colin C. The optimal mixed strategies are

Colin: $(\frac{2}{7}, 0, \frac{5}{7}, 0, 0)$ Rose: $(\frac{4}{7}, \frac{3}{7})$ (Do you see this on the graph?)

Here is the check:

Rose expectations against Colin optimal	Colin expectations against Rose optimal
Rose A: $\frac{2}{7}(-2) + \frac{5}{7}(1) = \frac{1}{7}$	Colin A: $\frac{4}{7}(-2) + \frac{3}{7}(3) = \frac{1}{7}$
Rose B: $\frac{2}{7}(3) + \frac{5}{7}(-1) = \frac{1}{7}$	Colin B: $\frac{4}{7}(5) + \frac{3}{7}(-3) = \frac{11}{7}$
	Colin C: $\frac{4}{7}(1) + \frac{3}{7}(-1) = \frac{1}{7}$
	Colin D: $\frac{4}{7}(0) + \frac{3}{7}(3) = \frac{9}{7}$
	Colin E: $\frac{4}{7}(-4) + \frac{3}{7}(8) = \frac{8}{7}$

The value of the game is $\frac{1}{7}$. The important thing in the check is that all of Colin's expectations against Rose optimal are $\geqslant \frac{1}{7}$: if even one had been less than $\frac{1}{7}$, we would know the claimed solution was in error.

What about games where both of the players have more than two strategy choices? First of all, they all do have solutions. The following theorem was proved by John von Neumann in 1928:

MINIMAX THEOREM. Every $m \times n$ matrix game has a solution. That is, there is a unique number v, called the *value of the game*, and there are optimal (pure or mixed) strategies for Rose and Colin such that

i) if Rose plays her optimal strategy, Rose's expected payoff will be $\geqslant v$, no matter what Colin does, and

ii) if Colin plays his optimal strategy, Rose's expected payoff will be $\leqslant v$, no matter what Rose does.

Furthermore, this solution can always be found as the solution to some $k \times k$ subgame of the original game.

In other words, the solution to an $m \times n$ game is always the solution to some 1×1 subgame (i.e., a saddle point), or some 2×2 subgame, or some 3×3 or larger square subgame. The pure strategies which are involved in the solution are called *active strategies*. They should be played according to certain probabilities, and the other strategies should not be played at all.

Here is an example of solving a 3×3 game by the method of equalizing expectations.

		Colin		
		A	B	C
	A	1	2	2
Rose	B	2	1	2
	C	2	2	0

Game 3.3

This game does not have a saddle point, and there is no dominance. We suppose that Colin plays Colin A, B, and C with probabilities $(x, y, 1 - x - y)$. Then Rose's expectations are

Rose A: $x(1) + y(2) + (1 - x - y)(2) = 2 - x$

Rose B: $x(2) + y(1) + (1 - x - y)(2) = 2 - y$

Rose C: $x(2) + y(2) + (1 - x - y)(0) = 2x + 2y.$

Rose will be unable to take advantage of Colin's mixed strategy if these expectations are all equal:

$$2 - x = 2 - y = 2x + 2y.$$

These equations give a system of two equations in two unknowns:

$$x - y = 0$$
$$2x + 3y = 2$$

which have the solution $x = y = \frac{2}{5}$. Thus Colin's expectation-equalizing mixed strategy is $(\frac{2}{5}, \frac{2}{5}, \frac{1}{5})$. By the symmetry of the matrix, this is also Rose's expectation-equalizing mixed strategy. We can check that these strategies do give a solution, and the value of the game is $\frac{8}{5}$.

It is important to realize that this method of equalizing expectations will fail if the solution to the 3×3 game involves a 1×1 or 2×2 subgame. You should check for a saddle point and dominance before you try equalizing expectations, and if equalizing expectations fails, you should be prepared to look for a 2×2 solution. The best way to do that is to use graphical analysis on the three possible 2×3 subgames.

With the techniques of this chapter, we can in principle solve any $m \times n$ matrix game. In practice, however, the work may be tedious. Consider, for example, solving a 4×4 game. We first check for saddle points, and then try to reduce the size of the game by dominance. If that doesn't work, we try to find a 4×4 solution by the method of equalizing expectations, which involves solving two sets of three equations in three unknowns. That will fail if the solution involves a 3×3 or 2×2 subgame, and we must then check those subgames. There are 16 possible 3×3 subgames to check, and 36 possible 2×2 subgames (though we could simplify the work by using the graphical technique on six 2×4 subgames). It may take awhile!

If you would like to practice solving matrix games by hand, I recommend [Williams, 1986], which has wonderful examples and many helpful hints. On the other hand, the most efficient method for solving large games is to use *linear programming* techniques. Most computer software packages which do linear programming have an option for solving matrix games, and I would recommend using one of them if you cannot solve a game within reasonable time.

If players don't know how to calculate mixed strategy solutions to matrix games, it is interesting to investigate how well they can play just intuitively. If you would like to experiment, here is a game to try.

		Colin					
		A	B	C	D	E	F
Rose	A	4	−4	3	2	−3	3
	B	−1	−1	−2	0	0	4
	C	−1	2	1	−1	2	−3

Game 3.4

Don't try to solve it. Just sit down and play it with a friend for 30 plays, keeping track of strategies chosen and payoffs. When you are finished, think about what you learned in the course of play. If you were Rose, can you guess which Rose strategy should be played most? If you were Colin, can you guess which Colin strategies are active in the solution?

When my Game Theory class played this game, the results were mixed. The Rose players learned the optimal Rose strategy quite efficiently:

	First ten games	Last ten games	Optimal	Expectation vs. Colin optimal
Rose A:	36%	21%	24% (10/42)	$-3/42 = -.07$
Rose B:	25%	29%	21% (9/42)	$-3/42 = -.07$
Rose C:	39%	50%	55% (23/42)	$-3/42 = -.07$

On the other hand, the Colin players did not learn very well:

	First ten	Last ten	Optimal	Expectation vs. Rose optimal	Comments
Colin A:	14%	9%	0%	$8/42 = .19$	slightly bad
Colin B:	36%	34%	36%	$-3/42 = -.07$	
Colin C:	8%	12%	0%	$35/42 = .83$	very bad
Colin D:	16%	18%	57%	$-3/42 = -.07$	
Colin E:	17%	17%	0%	$16/42 = .38$	dominated by B
Colin F:	9%	9%	7%	$-3/42 = -.07$	

The Colin players persisted in playing non-active, inferior strategies, playing them 38% of the time even in the last ten games of thirty. Most Colin players did not notice that Colin E is dominated by Colin B. Colin players did not play Colin D enough, in spite of the important role of Rose C in Rose's mixed strategy, and they actually increased plays of the highly unfavorable Colin C. The result was that although this game is slightly favorable to Colin (with a value of $-.07$), the Rose players ended up ahead, winning an average of $+.28$ per game in the last ten games. Perhaps the moral is that analysis can help—the solution to a game is sometimes not easily intuited on the basis of experience alone. On the other hand, Exercise 9 asks you to verify experimentally that *systematic* use of experience can lead to optimal play.

Exercises for Chapter 3

1. Consider a general 2×2 game:

Colin

Rose	A	B
A	a	b
B	c	d

The game will have a saddle point unless the two largest entries are diagonally opposite each other, so suppose the two largest entries are a and d. Suppose Colin plays A and B with probabilities x and $(1-x)$.

a) Show that the value of x which will equalize Rose's expectations for Rose A and Rose B is

$$x = \frac{d-b}{(a-c)+(d-b)}.$$

(This verifies Williams' oddments of $|b-d|$ and $|a-c|$.)

b) Show that the value of the game is

$$v = \frac{ad-bc}{(a-c)+(d-b)}.$$

2. Take a 2×2 game which has a saddle point and investigate what happens if you
   a) try to find a mixed strategy for Colin which will equalize Rose's expectations.
   b) use Williams' oddment method.

3. Solve the following games:

a)

	A	B
A	−3	5
B	−1	3
C	2	−2
D	3	−6

b)

	A	B
A	−2	5
B	1	2
C	0	−2
D	0	4

c)

	A	B	C	D	E
A	−4	2	0	3	−2
B	4	−1	0	−3	1

4. Some games have more than one solution. The value of the game is fixed, but the players may have several different strategies which ensure this value.
   a) Draw the graph for the following game. What happens?

Colin

Rose	A	B	C
A	−2	0	2
B	3	1	−1

   b) Show that there are two different optimal strategies for Colin, corresponding to the solutions for two different 2×2 subgames. The third 2×2 subgame does not yield a solution. In the graph, what is different about that subgame?

5. Solve the following games:

a)

	A	B	C
A	3	0	1
B	−1	2	2
C	1	0	−1

b)

	A	B	C
A	5	2	1
B	4	1	3
C	3	4	3
D	1	6	2

c)

	A	B	C	D
A	4	−3	2	−4
B	4	−4	4	−2
C	0	1	−3	2
D	−5	2	−7	2
E	3	−2	2	−2

6. We checked that the solution to the Game 3.4 is Rose $(\frac{10}{42}, \frac{9}{42}, \frac{23}{42})$, Colin $(0, \frac{15}{42}, 0, \frac{24}{42}, 0, \frac{3}{42})$, with value $-\frac{3}{42}$. Derive this solution by using the method of equalizing expectations on the 3×3 game Rose ABC vs. Colin BDF.

7. In a simplified version of the Italian finger game *Morra*, each of two players shows one finger or two fingers, and simultaneously guesses how many fingers the other player will show. If both players guess correctly, or both players guess incorrectly, there is no payoff. If just one player guesses correctly, that player wins a payoff equal to the total number of fingers shown by both players.
   a) Each player has four strategies. Write the 4×4 matrix for this game. (Note that it should have a kind of symmetry.)
   b) Since the game is symmetric, its value should be zero. Show that the strategy $\frac{5}{12}$ show two–guess one, $\frac{7}{12}$ show one–guess two is optimal for both players.

   (In actual Morra, players can show one, two or three fingers. For its solution, see [Williams, 1986], pages 163–165.)

8. Solve the following game, and verify your solution. (If you can guess the solution, it will save you work.)

		Colin			
		A	B	C	D
Rose	A	1	2	2	2
	B	2	1	2	2
	C	2	2	1	2
	D	2	2	2	0

9. Suppose Rose and Colin play many rounds of a game by the following procedure. In the first round, each player chooses a pure strategy arbitrarily, say Rose A and Colin A. In the $(n + 1)$st round, each player plays the pure strategy which has the best expected value against the mixture of strategies used by the other player in the first n rounds.
   a) Write a computer program to carry out this procedure for Game 3.4 You should find that the players' mixtures converge to their optimal mixed strategies.† How soon do Colin's non-active strategies stop being played?
   b) (Suggested by Peter Ungar) It would be dangerous to play strictly according to this procedure, because your opponent might realize what you are doing. Suppose in Game 3.4 that Colin plays by this procedure, but Rose realizes he is doing it and in each round plays her best response to what she knows Colin is going to do. Adapt your computer program to investigate the result. Poor Colin will do badly, but will the long-run mixtures still converge to the optimal mixed strategies?

†This is a celebrated theorem of Julia Robinson [1951].

4. Application to Anthropology: Jamaican Fishing

One important school of anthropological thought, known as *functionalism*, holds that customs, institutions or behavior patterns in a society can be interpreted as functional responses to problems which the society faces. One method, then, of understanding the organization of societies would be to identify problems and stresses, see what kinds of behavior would provide good solutions, and compare a society's behavior patterns to those solutions. For example, incest taboos can be interpreted as societal solutions to genetic problems caused by inbreeding.

In the 1950's some pioneering anthropologists began to use game-theoretic ideas in the service of functionalism. For example, Moore [1957] proposed that one could interpret *divination* as a societal mechanism for implementing mixed strategy solutions to a game. Recall that in a game with a mixed strategy solution, it is crucial that strategies be selected randomly, with certain probabilities. There must be no pattern of strategy choices, even an unconscious one, which could be noticed and exploited by an opponent. We suggested making choices by using a chance mechanism, but a society might feel uncomfortable making important decisions this way. A way to obtain the same goal would be to have a shaman read the favorable course of action from a randomly generated pattern of caribou bones. Interpretive rules might evolve to select different alternatives with optimal probabilities.

The first quantitative application of two-person game theory to an anthropological problem was Davenport [1960], a classic and still controversial paper on Jamaican fishing. Davenport studied a village of two hundred people on the south shore of Jamaica, whose inhabitants make their living by fishing. The fishing grounds extend outward from shore about 22 miles.

> "Twenty-six fishing crews in sailing, dugout canoes fish this area by setting fish pots, which are drawn and reset, weather and sea permitting, on three regular fishing days each week... The fishing grounds are divided into inside and outside banks. The inside banks lie from 5 to 15 miles offshore, while the outside banks all lie beyond... Because of special underwater contours and the location of one prominent headland, very strong currents set across the outside banks at frequent intervals in both easterly and westerly directions. These currents... are not related in any apparent way to weather and sea conditions of the local region. The inside banks are almost fully protected from the currents." [Davenport, 1960]

The captains of the canoes might conceivably adopt three different fishing strategies:

Inside: put all pots on the inside banks.

Outside: put all pots on the outside banks.

In-Out: put some pots on the inside banks, some pots on the outside.

These different fishing strategies have a number of different advantages and disadvantages. Here are some of them:

- Since travel times are longer, crews following the Outside or In-out strategies can set fewer pots.
- When the current is running, it is harmful to outside pots in a number of ways. The bamboo floats marking the location of the pots are dragged underwater and the fishermen cannot find them; the pots are moved around on the bottom and may be smashed; fish in the pots may be killed by changes in temperature and other conditions induced by the current.
- The outside banks produce higher quality fish, both in varieties and in size. In fact, if many outside fish are available, they may drive the inside fish off the market.
- The Outside or In-out strategies require sturdier canoes. Inside fishermen often buy their canoes used from fishermen who go outside. Since Outside canoes are newer and sturdier, their captains dominate the sport of canoe racing, which is prestigious and offers large purses.

Davenport gathered data to estimate the payoff for following each of the three possible strategies when the current is running, and when the current is not running:

		Current	
		Run	Not run
Fishermen	Inside	17.3	11.5
	Outside	−4.4	20.6
	In-Out	5.2	17.0

Game 4.1

The payoffs are average profits in English pounds per fishing month to the captain of the fishing canoe. Davenport notes that these estimates were made before he had any knowledge of game theory or plans to do a game-theoretic analysis, so they should be free of unconscious bias. However, knowing game theory, we can treat this as a 3×2 game, solve for the optimal fishing strategy and compare it to the actual fishing pattern of the villagers. The game does not have a saddle point and there is no dominance.[†] The diagram is shown in Figure 4.1. The lowest

[†]If you read Davenport's paper, you will notice that he had an incorrect understanding of the idea of dominance.

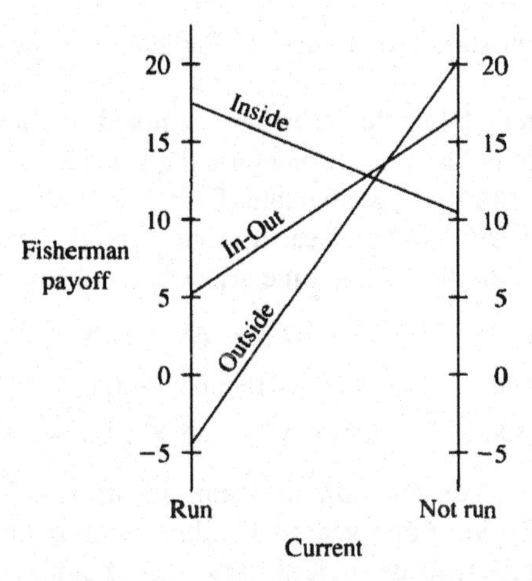

Figure 4.1 Payoff diagram for Game 4.1

point on the upper envelope involves the Inside and In-Out strategies. Solving that 2×2 game we get an optimal strategy of 67% Inside, 33% In-Out for the fishermen; an optimal strategy of 31% Run, 69% Not run for the current; and a value of 13.3.

The comparison of this game-theoretic solution with actual behavior is quite striking. First, no captains followed the Outside strategy, which the villagers characterized as entirely too risky. Second, in the period Davenport observed, 69% of the captains followed the Inside strategy and 31% followed the In-Out strategy. Third, even the current seemed to be following its optimal strategy fairly closely: estimates over a two year period were that it ran 25% of the time. The conclusion is that this society has adapted well to its natural and economic environment.

Davenport's analysis went unchallenged for several years and was widely cited in the anthropological literature. However, Kozelka [1969] and Read and Read [1970] independently pointed out that there is a serious flaw in the analysis. The fishermen's opponent in this game is a natural phenomenon, the current. It is not a reasoning entity, and its behavior is not affected by what the fishermen do. In particular, it would not adjust its behavior to take advantage of non-optimal play by the fishermen. The fishermen's correct behavior in this context would be to use the Expected Value Principle. They should observe the mixed strategy 25% Run, 75% Not run being followed by the current, calculate the expected values of their various strategies, and follow the strategy with the highest expected value. These expected values are

$$\begin{aligned} \text{Inside:} &\quad .25 \times 17.3 + .75 \times 11.5 = 12.95 \\ \text{Outside:} &\quad .25 \times (-4.4) + .75 \times 20.6 = 14.35 \\ \text{In-Out:} &\quad .25 \times 5.2 + .75 \times 17.0 = 14.05. \end{aligned}$$

All of the fishermen should fish outside! Perhaps this society is not so well-adapted after all.

The key to the rebuttal of this criticism, as noted in Bagnato [1974], is the villagers' explanation of why they do not use the Outside strategy: it is too risky. The current is notoriously unpredictable. Even if on average it may run 25% of the time, in the short run of a year it might run considerably more or less. Suppose one year it ran 35% of the time. The expected payoffs would be

$$\begin{aligned} \text{Inside:} \quad & .35 \times 17.3 + .65 \times 11.5 = 13.53 \\ \text{Outside:} \quad & .35 \times (-4.4) + .65 \times 20.6 = 11.85 \\ \text{In-Out:} \quad & .35 \times 5.2 + .65 \times 17.0 = 12.87. \end{aligned}$$

Suppose the village needs a certain minimum payoff, say 13 pounds per canoe per month, over the course of a year to avoid hardship or economic ruin. If the fishermen followed the Outside strategy the village could be ruined by a long run of bad behavior by the current. The advantage of the game theoretic minimax solution is that it guarantees an average income of at least 13.3 pounds per canoe per month regardless of what the current does. In this and similar situations it is this security property of the minimax solution which may make it desirable even when the opponent is not a reasoning entity. We will discuss this kind of "game against nature" at greater length in Chapter 10.

Exercises for Chapter 4

1. By the Expected Value Principle, Outside is the best strategy if the current is running 25% of the time, while Inside is best if the current is running 35% of the time. Find a current percentage for which In-Out would be the best fishing strategy.
2. Suppose the payoff for In-Out with no current were 15.0 instead of 17.0. What would be the effect on the game-theoretic solution? On the value of the game?

5. Application to Warfare: Guerrillas, Police, and Missiles

Zero-sum games represent conflict situations, and our solution theory for them prescribes rational strategies for conflict. Since the most extreme form of conflict is war, it is not surprising that some of the first proposed applications of game theory were to tactics in war. Haywood [1954] and Beresford and Peston [1955] describe some applications of game theory to situations from World War II. In this chapter we will consider two applications in more modern settings. Both will be far too simplified to be realistic, but they may give the flavor of what kinds of contributions zero-sum game theory might make to military tactics.

Our first example is abstracted from a situation which might arise in guerrilla warfare. I will call the game "Guerrillas vs. Police." There are m guerrillas, n police, and two government arsenals which the guerrillas would like to capture and the police must defend. The guerrillas will attack one or both arsenals, and they will capture any arsenal which they attack with a force stronger than the force of defenders. The guerrillas win the game if they capture even one arsenal (they then have arms to continue fighting); the police win only if they successfully defend both arsenals.

The guerrillas can clearly win if $m > n$: they just attack either arsenal with full force. The police win if $n \geqslant 2m$: they defend each arsenal with a force of at least m. The interesting cases are when $m \leqslant n < 2m$. To get a feeling for this situation, suppose $m = 2$ and $n = 3$. The guerrillas' choice is how to divide their force between the arsenals, 2-0 or 1-1. If they divide 2-0 they still have to decide which arsenal to attack, but the solution to this part of the problem is clear. They should decide randomly by flipping a coin, since any non-random choice might be anticipated by the police. The police must decide whether to divide 3-0 or 2-1, and then flip a coin to decide where to send the stronger force. Here is the payoff matrix, with payoffs to the guerrillas (1 for a win, 0 for a loss):

		3 Police	
		3-0	2-1
2 Guerrillas	2-0	$\frac{1}{2}$	$\frac{1}{2}$
	1-1	1	0

Game 5.1

The $\frac{1}{2}$'s are expected values. If the guerrillas split 2-0, half of the time they will attack the arsenal which has the stronger defending force, and lose. The other half of the time they will attack the weaker arsenal and win. The game has a saddle point with value $\frac{1}{2}$.

For some other force strengths, the optimal strategies may be mixed:

		4 Police		
		4-0	3-1	2-2
	4-0	$\frac{1}{2}$	1	1
4 Guerrillas	3-1	1	$\frac{1}{2}$	1
	2-2	1	1	0

Game 5.2

		9 Police				
		9-0	8-1	7-2	6-3	5-4
	7-0	$\frac{1}{2}$	$\frac{1}{2}$	$\frac{1}{2}$	1	1
7 Guerrillas	6-1	1	$\frac{1}{2}$	$\frac{1}{2}$	$\frac{1}{2}$	1
	5-2	1	1	$\frac{1}{2}$	$\frac{1}{2}$	$\frac{1}{2}$
	4-3	1	1	1	$\frac{1}{2}$	0

Game 5.3

We have already solved these games in Chapters 2 and 3, although they appeared there with their payoffs multiplied by two. The solutions are

for Game 5.2:	Guerrillas	$(\frac{2}{5}, \frac{2}{5}, \frac{1}{5})$
	Police	$(\frac{2}{5}, \frac{2}{5}, \frac{1}{5})$
		Value $= \frac{4}{5}$
for Game 5.3:	Guerrillas	$(\frac{2}{3}, 0, 0, \frac{1}{3})$
	Police	$(0, 0, \frac{2}{3}, 0, \frac{1}{3})$
		Value $= \frac{2}{3}$

In Game 5.2 all strategies are active. In Game 5.3 each side has just two active strategies. One is the "maximum possible or necessary force" split (7-0 for the guerrillas, 7-2 for the police), and the other is the "equal" split (4-3 for the guerrillas, 5-4 for the police). The former should be used twice as often as the latter.

My game theory classes worked out the solutions for Guerrillas vs. Police for many small values of m and n. The values of these games to the guerrillas are shown in Table 5.1. There are a number of interesting patterns. All of the games with value $\frac{1}{2}$ are saddle point games, where the guerrillas and police both do a maximum necessary force split. All of the games down the diagonal are all-strategies-active games similar to Game 5.2. All of the games with value $\frac{2}{3}$ have two active strategies like Game 5.3. Do you see other patterns? One interesting qualitative conclusion is that there are broad ranges over which increasing the number of police does not decrease the probability that the guerrillas will win. For instance, the value of the game is $\frac{1}{2}$ over the whole range $\frac{3}{2}m \leq n < 2m$.

Number of Guerrillas (m) \ Number of Police (n)	1	2	3	4	5	6	7	8	9	10	11	12
1	$\frac{1}{2}$											
2		$\frac{2}{3}$	$\frac{1}{2}$									
3			$\frac{3}{4}$	$\frac{1}{2}$	$\frac{1}{2}$				all 0's			
4				$\frac{4}{5}$	$\frac{2}{3}$	$\frac{1}{2}$	$\frac{1}{2}$					
5					$\frac{5}{6}$	$\frac{2}{3}$	$\frac{1}{2}$	$\frac{1}{2}$	$\frac{1}{2}$			
6						$\frac{6}{7}$	$\frac{3}{4}$	$\frac{2}{3}$	$\frac{1}{2}$	$\frac{1}{2}$	$\frac{1}{2}$	
7							$\frac{7}{8}$	$\frac{3}{4}$	$\frac{2}{3}$	$\frac{1}{2}$	$\frac{1}{2}$	$\frac{1}{2}$
8								$\frac{8}{9}$	$\frac{4}{5}$	$\frac{2}{3}$	$\frac{2}{3}$	$\frac{1}{2}$
9									$\frac{9}{10}$	$\frac{4}{5}$	$\frac{3}{4}$	$\frac{2}{3}$
10					all 1's					$\frac{10}{11}$	$\frac{5}{6}$	$\frac{3}{4}$
11											$\frac{11}{12}$	$\frac{5}{6}$
12												$\frac{12}{13}$

Table 5.1 Values of Guerrillas vs. Police games for small m and n.

For a second kind of tactical game, we will consider a missile penetration problem from Johnson [1966], applicable to missile defense programs like the "star wars" program of the United States in the late 1980's. We'll call our countries Red and Blue. Suppose that Red wishes to destroy a Blue military base. Red has four missiles which will be fired in sequence. Two of the missiles have real warheads, while two are dummies. For defense, Blue has two anti-missiles. Each anti-missile can scan two Red missiles and destroy the first one it sees which has a real warhead.

Red must choose the order in which to send the live warheads and the dummies. We will use a notation in which "DWWD" means to send a dummy, warhead, warhead, dummy in that order. Blue's choice is when to fire the anti-missiles. Our notation will be that "13" means fire at the first and third Red missiles. Blue wins the game (payoff $+1$) if Blue destroys both Red warheads; Blue loses (payoff 0) if even one Red warhead gets through. To illustrate how the payoffs are calculated, consider Blue 12 against Red DWWD. The first Blue anti-missile sees Red missiles #1 and #2, and it kills #2, which has a live warhead. The second Blue anti-missile can scan Red missiles #2 and #3. Red #2 has already been destroyed, so this anti-missile kills Red #3. Blue wins, for a payoff of 1. The complete matrix is

		Red					
		WWDD	WDWD	WDDW	DWWD	DWDW	DDWW
Blue	12	1	1	0	1	0	0
	13	0	1	1	1	1	0
	14	0	0	1	0	1	0
	23	0	0	0	1	1	1
	24	0	0	0	0	1	1
	34	0	0	0	0	0	1

Game 5.4

First notice that Blue 14 is dominated by Blue 13, and Blue 24 and 34 are dominated by Blue 23. Blue should never hold an anti-missile to fire at the fourth Red missile, since this wastes the anti-missile's capability of scanning two Red missiles. Secondary dominance then rules out Red WDWD, DWWD and DWDW. The reduced game is

		Red		
		WWDD	WDDW	DDWW
Blue	12	1	0	0
	13	0	1	0
	23	0	0	1

The solution is the mixed strategy $(\frac{1}{3}, \frac{1}{3}, \frac{1}{3})$ for both Blue and Red, and the value of the game is $\frac{1}{3}$. Notice what Red's strategy means: Red should always fire its dummies together in a string, giving Blue a chance to waste an anti-missile. Notice also that Blue has only a $\frac{1}{3}$ chance of saving its base. One of the issues in any debate about missile defense systems is whether it would be easy to overwhelm such a system by dummy missiles.

Our two examples have had to do with small scale tactical situations in war, and I think they illustrate that zero-sum game theory could be useful in such situations. We have not discussed larger scale strategic questions. The reason is important. It is that above the level of small scale tactics most international conflict situations, even in war, cannot be reasonably modeled as zero-sum games. It is hardly ever true in international conflicts that interests are strictly opposed. There are usually outcomes which would be harmful to both sides. For example, poison gas was not used in World War II because both sides realized that its use would be mutually harmful. In the modern world, with the availability of nuclear weapons, war has moved very far from being a zero-sum game.

Exercises for Chapter 5

1. a) Set up and solve the guerrilla vs. police game with 6 guerrillas and 7 police, verifying that entry in Table 5.1.
   b) Can you see any patterns not mentioned in the text for the general problem of m guerrillas, n police? Can you show that any of the patterns you see do in fact hold for all appropriate values of m and n?

2. Set up and solve the missile game if Red has one warhead and three dummies and Blue has just one anti-missile.

3. Our Guerrillas vs. Police is a type of game known in the literature as a "Colonel Blotto" game. In the original paper introducing this type of game [McDonald and Tukey, 1949], Colonel Blotto had four units with which to battle for four forts against an enemy with three units. A battle would always be won by the side with the larger force, and the smaller force would be destroyed. Blotto's payoff was $+1$ for each enemy unit destroyed and each fort he captured, -1 for each of his units destroyed and each fort captured by the enemy. The game matrix is

		Enemy splits		
		3000	2100	1110
	4000	1	$\frac{1}{4}$	$-\frac{1}{2}$
	3100	$\frac{1}{2}$	$\frac{3}{4}$	$\frac{1}{2}$
Blotto splits	2200	$-\frac{1}{2}$	1	2
	2110	$\frac{1}{4}$	$\frac{1}{2}$	$\frac{3}{2}$
	1111	1	0	1

   a) Once a player has decided how to split his forces, he should certainly choose forts at random to send them to, so each payoff in this game is an expected value. One of the hardest-to-calculate payoffs is 3100 vs. 2100, where there are seven cases to consider. Do the calculation to verify this payoff.
   b) There is a 3×3 solution to this game, involving 3100, 2200 and 1111. Find it. There is something which looks a little odd about this solution, so be sure to check the solution in the full game.

4. Set up and solve a guerrillas vs. police game if there are *three* arsenals and
   a) 2 guerrillas, 4 police
   b) 3 guerrillas, 4 police.

6. Application to Philosophy: Newcomb's Problem and Free Will

One of the most persistent problems of philosophy is the problem of *free will.* Is human will free, or are our actions determined? One way to approach this problem is to consider the possibility of predicting any person's decision in some unconstrained choice situation. Suppose I ask you to decide consciously to hold up your left hand or your right hand, and tell you that I have written on a slip of paper my prediction of what you will do. Is it possible, in principle, for me to know enough about you to make my prediction with better than chance accuracy? Could God make such a prediction? If human actions are determined, an accurate prediction might be possible. If human will is free, your free will could thwart my, or God's, prediction.

In 1960, William Newcomb, a physicist at the Livermore Radiation Laboratories in California, posed a choice problem which can be phrased in game-theoretic terms, and which casts interesting light on the problem of free will. My description of it is based on a presentation by philosopher Robert Nozick [1969]. Nozick writes, and I agree, "It is a beautiful problem. I wish it were mine."

Suppose there are two black boxes which you cannot see into. Box #1 contains $1000. Box #2 contains either $1,000,000 or nothing, depending on something we will mention in a moment. You have two choices:

i) you may take both boxes, or

ii) you may take only Box #2.

Yesterday, a Being who you believe has superior predictive powers, made a prediction about what you will do today. If he (or she) predicted that you will take both boxes, he left the second box empty. If he predicted that you will take only Box #2, he put $1,000,000 in that box. (If he predicted that you will use a random device to make your choice, he left the second box empty.)

We need not be specific about the nature of the Being. It might be God, if God would stoop to playing such games. It might be an extra-terrestrial creature with superior mental powers, or one come through a time warp from the day after tomorrow. It might even be a psychologist who has put you through a battery of tests. The crucial thing is that you have reason to believe that this Being can make predictions with good accuracy. He might not be perfect, but he is right, say, 90% of the time.

Think about it. Which choice would you make? Both boxes, or only the second box?

Perhaps we can clarify the nature of this choice problem by thinking of it as a game between you and the Being:

		Being	
		Predict you will take both boxes	Predict you will take only Box #2
You	Take both boxes	$1000	$1,001,000
	Take only Box #2	0	$1,000,000

The Being has made his move, although you do not know what it is. Now it is your turn. Which strategy do you choose?

The problem here is that it is possible to give quite powerful arguments for choosing *either one* of the two possible courses of action:

ARGUMENT 1. "Suppose I take both boxes. Then the Being will almost certainly have predicted this and will have left Box #2 empty, so I will almost certainly get $1000. On the other hand, suppose I take only Box #2. Then the Being will almost certainly have predicted this and put $1,000,000 in Box #2, so I will almost certainly get $1,000,000. I would rather have $1,000,000 than $1000. I should take only Box #2."

ARGUMENT 2. "The Being made his prediction yesterday, and the $1,000,000 is either in Box #2, or it isn't. What I do today will not change that. If the $1,000,000 is there, it won't vanish just because I take both boxes, and I am better off taking both boxes and getting the extra $1000. It's not that I'm greedy, but why forego $1000? If the $1,000,000 is not there, I am certainly better off taking both boxes and getting $1000 rather than nothing. So in either case, I should take both boxes."

These arguments can be put in game-theoretic terms. Argument 1 is an argument from the Expected Value Principle. Suppose you believe that the Being predicts correctly with probability .9. Then the expected values of your two strategies are

$$\text{Take both boxes:} \quad .9 \times 1000 + .1 \times 1{,}001{,}000 = 101{,}000$$
$$\text{Take only Box \#2:} \quad .1 \times 0 + .9 \times 1{,}000{,}000 = 900{,}000.$$

You should take only Box #2. In fact, this continues to be true as long as the probability that the Being predicts correctly is larger than .5005.

On the other hand, Argument 2 is an argument from the Dominance Principle. Taking both boxes dominates taking only Box #2, and you should play the dominant strategy.

Under normal circumstances these two basic game theoretic principles would not conflict. You would estimate some probability x that the Being would predict you take both boxes, and the expected value computation would look like

Take both boxes: $1000x + 1{,}001{,}000(1 - x) = 1{,}001{,}000 - 1{,}000{,}000x$

Take only Box #2: $0x + 1{,}000{,}000(1 - x) = 1{,}000{,}000 - 1{,}000{,}000x.$

The expected value of the first strategy would be larger than the expected value of the second strategy for any value of x, and the Expected Value Principle would agree with the Dominance Principle. However, we are not dealing with normal circumstances here. The difference is your confidence that there is some connection between your choice and the Being's prediction, and it is this connection which produces the conflict.

Since your confidence in the Being's predictive power produces the dilemma, it is worthwhile thinking about how such confidence might be produced. One way might be by faith. Another might be by evidence. Suppose you have watched this experiment two hundred times, conducted on people like yourself. You know the people well enough to be satisfied they are not in collusion with the Being, and you have been checking the boxes to make sure there has been no dirty work. The results have been

		Being predicts subject will take	
		both boxes	only Box #2
Subject takes	both	90	10
	only Box #2	10	90

The Being has been right in 90% of the cases, giving the probabilities we used for the expected value calculation. Maybe you have been making bets that the Being will be correct, and winning 90% of the time. Now it is your turn. You have no reason to believe that you are very different from your friends. You would bet with pretty high odds that the Being will be correct again. It would not make sense to give up a 90% sure \$1,000,000. As Bar-Hillel and Margalit [1972] recommend, come "join the millionaires club!"

Against this strengthening of Argument 1 on the basis of evidence, Nozick offers the following strengthening of Argument 2. Suppose that Box #1 is transparent and you can see \$1000 in it. The front of Box #2 is opaque, but the back is transparent. Wouldn't it be nice if you had a dependable friend in back who could see the contents of both boxes and tell you what to do? But wait a minute—you don't need the friend, because you *know* what she would be urging. If Box #2 is empty she would be signaling frantically "Take both!" If Box #2 has \$1,000,000 in it she would be signaling, perhaps less frantically, "Take both!" Why should you ignore her and take only Box #2?

Nozick comments at this point:

> I have put this problem to a large number of people, both friends and students in class. To almost everyone it is perfectly clear and obvious what should be done. The difficulty is that these people seem to divide almost evenly

> on the problem, with large numbers thinking that the opposing half is just being silly.
>
> Given two such compelling opposing arguments, it will not do to rest content with one's belief that one knows what to do. Nor will it do to just repeat one of the arguments, loudly and slowly. One must also disarm the opposing argument: explain away its force while showing it due respect. [Nozick, 1969]

Nozick is not alone in finding people split on what to do in this situation. When Martin Gardner wrote about the problem in his *Scientific American* "Mathematical Games" column [1973], he invited readers to send in their solutions and suggestions. Of the 126 letters proposing solutions, 89 said they would take only Box #2, and 37 said they would take both boxes. An additional 18 respondents believed that the conditions of the problem were impossible to satisfy, a position we will examine shortly. In my game theory classes, belief in the Dominance Principle has been stronger and taking both boxes has been preferred by about three to two.

There is a connection between beliefs on the free will question and choices in Newcomb's problem. The stronger your belief that your will is free, the more likely you might be to resist the idea that a Being could predict your choice, and hence the more likely you might be to take both boxes. The connection, of course, is not absolute. Nozick and other philosophers have pointed out that the relationship between determinism and predictability of actions is subtle, with neither logically implying the other. However, many of Gardner's respondents who took both boxes specifically defended their choice as an affirmation of free will. It would be unfortunate, though, if a strong belief in free will determined the choice of taking both boxes, while less strong belief determined taking only Box #2. For it would be easy for a Being to know this, and weak believers could be rewarded by becoming millionaires...

There are two possible relationships between the free will question and Newcomb's problem which are more subtle. One is that strong belief in free will might produce the belief that the conditions of Newcomb's problem are impossible to fulfill. If human behavior in a choice situation is absolutely free and hence unpredictable, one can argue that Newcomb's Being could not exist and hence the problem does not exist. This is a comforting escape from the problem, but it is an extreme position. Remember that the payoffs in Newcomb's problem are such that it is advantageous to choose only Box #2 if the Being has even a 51% chance of predicting your choice correctly. Are we willing to deny even that modest amount of predictability to human actions?

The second possible relationship is even more extreme. One can argue, as Martin Gardner did, that Newcomb's paradox *proves* that human will is free, or at least that human actions in free-choice situations cannot, even in principle, be predicted with greater than chance accuracy. This argument sees the dilemma of Newcomb's problem as so strong that its existence is logically unacceptable. If a Newcomb Being could exist, it would produce the possibility of this dilemma. Therefore a Newcomb Being cannot exist and human choice is inherently unpredictable. I admire the audacity of this argument, and I like its conclusion, but I have difficulty granting Newcomb's problem the status of a logical impossibility.

Where does this leave us? Not with a solution, I think, unless you are sure that you have one—that you believe one argument and can disarm the other more effectively than by repeating your favored argument "loudly and slowly." We haven't decided the free will question either. However, I hope we have shown that a game-theoretic setting can give a new and original way of looking at a very old and important philosophical problem.

By the way, Nozick recommends (somewhat half-heartedly) taking both boxes. So does Isaac Levi [1975], on the basis that it is the cautious maximin strategy. Newcomb himself, Bar-Hillel and Margalit, and Steven Brams [1975] have recommended taking only Box #2. I would take both boxes, and prepare to bear my envy of the millionaires who made the other choice.

Exercises for Chapter 6

1. Levi [1975] asks us to consider a modified situation in which if you take both boxes and Box #2 happens to contain $1,000,000 (because the Being made a mistake), you have to pay a "greediness fee" of $1500.
   a) Write out the game corresponding to this situation, and show that the Dominance Principle no longer applies.
   b) Can you think of reasons which might lead you to take both boxes in this situation?
2. John Ferejohn, reported in Brams [1975], noted that Newcomb's dilemma would vanish if we thought of the situation as a "game against Nature" rather than a game against the Being. After all, the Being has already made his move and is no longer an active player, so what matters is which of two "states of Nature" holds. Ferejohn takes these states of nature to be "Being predicted correctly" and "Being predicted incorrectly," getting the following decision problem:

		State of Nature	
		Being correct	Being incorrect
You	take both boxes	$1000	$1,001,000
	take only Box #2	$1,000,000	0

   a) Note that the Dominance Principle doesn't apply here. What advice does the Expected Value Principle give you?
   b) Can you think of reasons to doubt the correctness of this resolution of the dilemma? We will discuss games against Nature in Chapter 10.
3. Consider the argument that it is not possible to accumulate convincing evidence that the Being can predict correctly by watching him predict correctly for other people "like you." The idea is that in order for the person before you to be like you, she would have had to see as much evidence as you have seen, and this would lead to an infinite regress—the Being would have to have been making predictions forever. Do you believe that this argument shows that Newcomb's situation is impossible, thereby resolving the dilemma?

7. Game Trees

In matrix games we have assumed that the players make their choice of strategy simultaneously, without knowledge of what the other player is choosing. This would seem to be a major limitation of the theory, since in real conflict situations decisions are often made sequentially, with information about previous choices becoming available to the players as the situation develops. In this chapter we will consider a method of modeling such sequential choice situations by a *game tree*. We will find that, perhaps surprisingly, this new model can always be reduced to our old model of a matrix game.

As a first example, consider the following radically simplified version of a game of poker. Each of two players, Rose and Colin, puts $1 into the pot as "ante." Each is then dealt a hand, which consists of one card, from a large deck which consists only of aces and kings. Colin must then decide whether to *bet* $2 or to *drop*. If he drops, Rose wins the pot. If Colin bets, Rose must decide whether to *call* by matching Colin's bet, or to *fold*. If she folds, Colin wins the pot. If Rose calls, the players compare their hands and the higher card wins the pot. If the hands tie, the pot is split equally.

To model this game, notice that it is a sequence of choices. The first choice is made by CHANCE, which deals the hands. CHANCE has four choices: ace to both (A, A), ace to Rose and king to Colin (A, K), vice versa (K, A), and king to both (K, K). We assume that each of these pairs of hands is dealt with probability $\frac{1}{4}$. Following CHANCE's choice, Colin decides to bet or drop. If Colin bets, Rose decides to call or fold. At the end, payoffs are assigned to both players. Figure 7.1 shows the sequence of choices as a game tree, in which

i) Each inside node is labeled by the player (including CHANCE) who makes a choice at that node.
ii) Each branch leading downward from a node represents a possible choice made by the player at that node.
iii) Each branch corresponding to a choice made by CHANCE is labeled by the probability that CHANCE will make that choice.
iv) Each final node is labeled by payoffs to the players.

Such a game tree could also be used to describe situations in which the players make several choices in sequence. For instance, the next most complicated model of poker would give Rose a choice of call, fold or raise. If Rose raised, Colin would have another choice of whether to call or fold. A game tree can also describe games with more than two players.

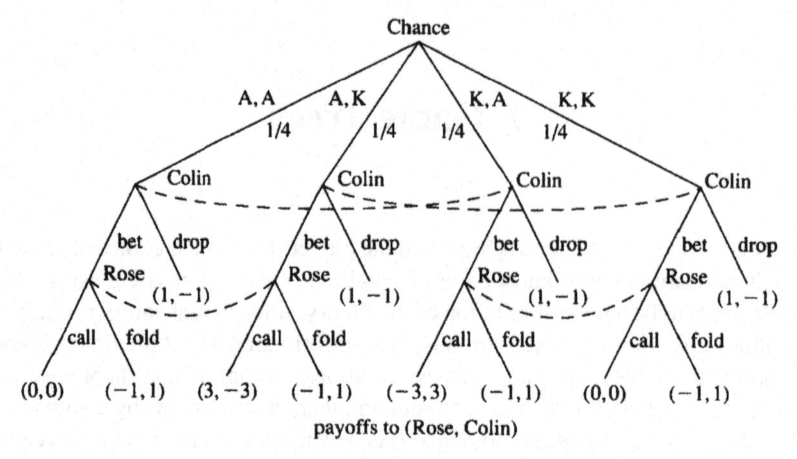

Figure 7.1

The dotted curves in Figure 7.1 represent one further, very important aspect of the game. When it comes time to make a choice, a player does not always know where he is on the game tree. For example, when deciding whether to bet or drop, Colin knows his own card, but he does not know Rose's card. If his card is an ace, Colin knows he is at the first or third node in the second row, but he doesn't know which. Hence we say that these two nodes are in the same *information set* and join them by a dotted curve. The other dotted curves join nodes in other information sets. In a general game tree we thus also have

v) The nodes for each player are partitioned into information sets. When a player makes a choice, he knows he is at a node in a particular information set, but he doesn't know which node.

vi) Nodes in the same information set have the same number of branches leading from them, labeled in the same way.

If condition vi) did not hold, a player could know where he was in the information set by which choices were available to him.

I said that any two-person game which can be represented by a game tree can in fact be represented by a game matrix. The connecting idea is the idea of a *strategy*. A strategy in a game tree is a complete description by a player of what choice he will make at any information set in which he might find himself during the course of the game. For instance, in our poker game a strategy for Colin might be "In my first information set (i.e., if I have an ace) I will bet; in my second information set (i.e., if I have a king) I will drop." Since Colin has two possible choices in each of two information sets, he has $2 \times 2 = 4$ possible strategies. In this game, Rose also has four possible strategies. If we know which strategy each player will use, we can trace the resulting course of play through the game tree, except for moves by CHANCE. Since we know probabilities for moves by CHANCE, we can calculate the expected values of payoffs to the players. If we label the rows and columns of a matrix by the possible strategies of Rose and Colin, and enter the corresponding expected payoffs in the matrix, we have a matrix game which corresponds to the game originally presented as a tree. Thus every game tree can be reduced to a game matrix.†

To illustrate, suppose that in the poker game Colin chooses the strategy of betting only if he has an ace, and Rose chooses the strategy of calling only if she has an ace. We can then calculate the outcome for each possible CHANCE move:

Probability	Rose hand, Colin hand	Outcome	Payoff to Rose
$\frac{1}{4}$	A, A	Colin bets, Rose calls, tie	0
$\frac{1}{4}$	A, K	Colin drops	+1
$\frac{1}{4}$	K, A	Colin bets, Rose folds	−1
$\frac{1}{4}$	K, K	Colin drops	+1

The expected payoff to Rose for these choices of strategies is $(\frac{1}{4})(0) + (\frac{1}{4})(1) + (\frac{1}{4})(-1) + (\frac{1}{4})(1) = \frac{1}{4}$. If we do the corresponding calculation for the other fifteen possible pairs of strategies and enter the results in a matrix, we get the matrix game

		Colin bets			
		always	A only	K only	never
Rose calls	always	0	$-\frac{1}{4}$	$\frac{5}{4}$	1
	A only	$\frac{1}{4}$	$\frac{1}{4}$	1	1
	K only	$-\frac{5}{4}$	$-\frac{1}{2}$	$\frac{1}{4}$	1
	never	−1	0	0	1

Game 7.1

From the matrix form, we see that the game has two saddle points, with value $\frac{1}{4}$. Colin should always bet with an ace, and may bet with a king. Rose should call only with an ace. Rose can expect to win an average of 25 cents per game.

For a different kind of example of modeling a sequential choice situation with a game tree and then reducing the tree to a game matrix, consider a very over-simplified description of the Cuban missile crisis between the United States under John Kennedy and the Soviet Union under Nikita Khrushchev in 1963. Khrushchev starts the game by deciding whether or not to place intermediate range ballistic missiles in Cuba. If he does place the missiles, Kennedy has three options: do nothing, blockade Cuba, or eliminate the missiles by a surgical airstrike. If Kennedy chooses the aggressive action of a blockade or an airstrike,

†In the technical literature a game tree is called a *game in extensive form*, and a game matrix is called a *game in normal form*. The theorem then states: any game in extensive form can be reduced to a game in normal form.

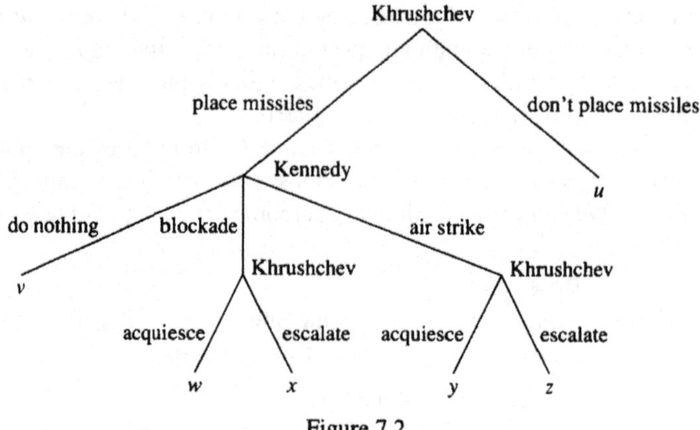

Figure 7.2

Khrushchev may acquiesce, or he may order escalation, possibly leading to nuclear war.

Figure 7.2 shows the game tree. I have represented the outcomes by letters, which we would have to replace by payoffs to Kennedy and Khrushchev to complete the description. I will not try to do that, but we should certainly note that the game is not zero-sum: x and z are certainly worse than u for both players. Since the players know each others' previous moves, each information set consists of just a single node.

In this game, Kennedy has three strategies, corresponding to his three choices at his single node. Khrushchev's strategies are more complicated. If Khrushchev chooses at the first node not to place missiles, the game is done. If he chooses to place missiles, the game could lead to either of his other nodes, and he must specify what he would do at each of them. Hence Khrushchev has a total of five strategies:

A. Don't place missiles.

B. Place missiles. Then always acquiesce.

C. Place missiles. Then acquiesce to a blockade, escalate against an airstrike.

D. Place missiles. Then escalate against a blockade, acquiesce to an airstrike.

E. Place missiles. Then escalate against any aggressive action.

The resulting matrix game would be

		Khrushchev				
		A	B	C	D	E
Kennedy	A	u	v	v	v	v
	B	u	w	w	x	x
	C	u	y	z	y	z

Game 7.2

Although any game tree can be reduced to a matrix game, there is a practical problem, which is that the number of possible strategies may be astronomically large, even for relatively small game trees. For example, consider just the first two moves of a game of chess. There are 20 possible first moves by White (16 pawn moves and 4 knight moves). Each of these may be responded to by 20 possible Black moves. Hence if we agreed to terminate the game at this stage, the game tree would have 400 branches. White would have just 20 strategies. However, since a strategy for Black would have to specify which of 20 choices to make for each of White's 20 possible choices, Black has $20^{20} \approx 10^{26}$ possible strategies. We could, with a little patience, write out a game tree with 400 branches, but no amount of patience would enable us to write out a matrix with 100 septillion columns! In practical situations it may be preferable to work directly with the game tree instead of trying to write out the corresponding matrix.

A game is said to have *perfect information* if in the game tree

- no nodes are labeled by CHANCE, and
- all information sets consist of a single node.

In other words, chance plays no part in the game, and each player always knows all previous moves which have been made in the game. Games of perfect information can be analyzed by a technique known as *truncation*, or *tree pruning*. Figure 7.3 shows a zero-sum example. In the original tree, consider Colin's last choice. Colin would always do best to choose the branch leading to the smallest Rose payoff, so we can truncate the tree by removing its final branches and labeling each new final node by the smallest Rose payoff below that node. This gives the second tree in the Figure. For the final choices in that tree, Rose would always do best to choose the branch leading to the largest Rose payoff, so we can truncate again, and get the final tree. In this tree, it is clear that Colin should choose the rightmost branch. Hence the rational outcome of the game is a payoff of -2 to Rose, and by tracing the truncation steps backwards we can find the sequence of choices leading to this outcome.

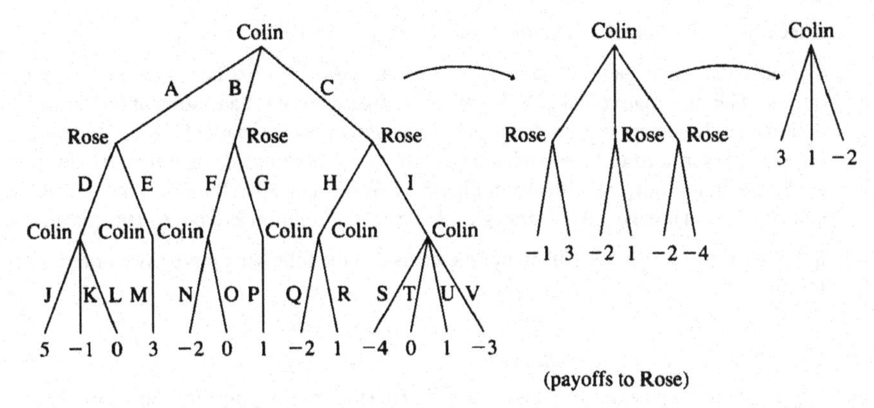

Figure 7.3

If we wrote the matrix form for this game (Rose has 8 strategies, Colin has 13—Exercise 4) and followed the tree truncation process in the matrix, we would find that the first step eliminates 10 Colin strategies by dominance. The second step eliminates 7 Rose strategies by secondary dominance. The third step eliminates 2 more Colin strategies by tertiary dominance, leaving a saddle point with value -2. Since this process would work for any two-person zero-sum game of perfect information, any such game will have a saddle point, and the saddle point strategies can be found by iterated dominance.

One interesting consequence of this analysis is a theorem first proved by the mathematical logician Ernst Zermelo [1912]. Consider a finite game of complete information between two players White and Black in which there are just two possible outcomes, "White wins" or "Black wins." Then exactly one of the following two statements must hold:

- White has a strategy which can force a win for White, no matter what Black does, or
- Black has a strategy which can force a win for Black, no matter what White does.

Notice that it is by no means trivial to go from the statement that every play of the game results in either a win for White or a win for Black, to the statement that one or the other player has a strategy which can force a win regardless of what the other player does. Yet this follows from the truncation technique. Think of the game as zero-sum with payoffs of White of $+1$ for a win, -1 for a loss. Truncation shows that the game has a saddle point, with a value of either $+1$ or -1. In the first case White's saddle point strategy forces a win for White; in the second case Black's saddle point strategy forces a win for Black. Exercise 7 asks you to consider what happens when draws are possible. Straffin [1985] gives an extension of Zermelo's theorem to games with three players.

Exercises for Chapter 7

1. Check two of the payoffs in Game 7.1: a) $\frac{5}{4}$ b) $-\frac{1}{2}$.
2. Consider the same game of poker, except that the deck now has aces, kings, and queens. The new matrix is 8×8. If you are energetic, write it and solve it by iterated dominance. If you are less energetic, you may take my word that only three strategies for each player need to be considered (the others are dominated): betting (or calling) always, with ace only, or with ace or king only. Write and solve the 3×3 matrix. The solution should involve mixed strategies. Interpret the solution as advice to the players.
3. Suppose the players in the Cuban missile crisis had the following preference orderings for the outcomes.

   Kennedy: w, y, u, v, x, z.

   Khrushchev: v, u, w, y, z, x.

   Use truncation on the tree in Figure 7.2 to find the rational outcome for this game. That is not what happened. Can you think of some possible explanations for why not?

4. Explain why in the game of Figure 7.3, Rose has 8 strategies and Colin has 13.

5. For the game in Figure 7.4
   a) Find the solution by truncation.
   b) List all Colin strategies. Be sure to describe each one clearly.
   c) List all Rose strategies.
   d) Write the matrix for the game.
   e) Solve the matrix and verify that the solution is the same as you found by tree truncation.

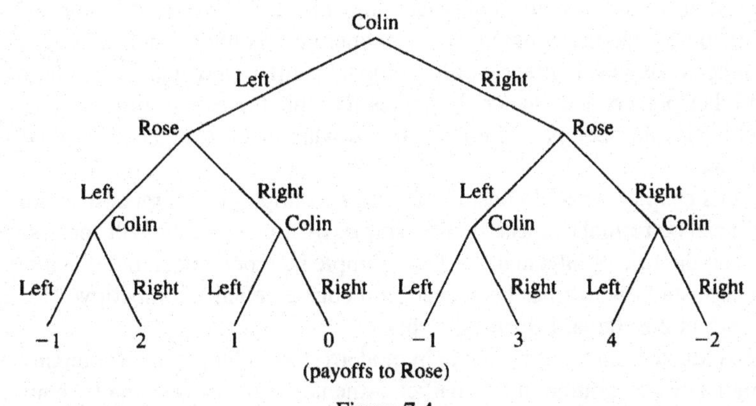

Figure 7.4

6. Suppose that the game in Figure 7.4 is modified so that Colin does not know what choice Rose made before Colin has to make his second choice.
   a) Show the information sets given by this condition.
   b) List all Colin strategies (there will be fewer).
   c) Write the game matrix.
   d) Solve the game. The solution is 3×3, with a value of $\frac{1}{2}$. Has Colin's loss of information helped or hurt Colin?

7. What can you say about finite White-Black games of perfect information where there are three possible outcomes: White wins, Black wins, or draw?

8. Application to Business: Competitive Decision Making

In the business world, companies often have to make decisions which involve strategic uncertainty about what other companies will do. Such decisions usually also involve uncertainty about future economic conditions, market size, costs, and other variables. In other words, companies often play games which involve both other players and chance. In games like this the role of information, both about what other companies might do and what chance might do, can be very important.

In this chapter we will consider a simple example of a competitive situation which can be formulated as a two-person zero-sum game. We will focus specifically on the role of information. The example is hypothetical, but it is modeled on a case study which has been used in a course called "Competitive Decision Making" at the Harvard Business School.†

Zeus Music is an industry leader in modern stereo equipment, a company with strong name recognition in the market. Athena Acoustics is a smaller company, well known for its high quality research and development work. Both companies have been developing a promising new "hexaphonic" sound system. In this revolutionary concept of a total music environment, the listener sits suspended in mid-air while six speakers project music from left, right, front, back, below and above. The effect is said to be extraordinary. It is at this point uncertain how large the market for such a system will be. It might be small, with potential profits of about $24 million yearly, or it might be large, with profits about $40 million. Analysts at Zeus estimate the chances for a small vs. large market at about 50-50.

Zeus and Athena must each decide what kind of product to produce for this market—a highest quality system to appeal to afficionados, or a slightly lower quality system with more gadgetry to appeal to a market with less sensitive musical discrimination. If the market is small, the high quality product will sell better, whereas if the market turns out to be large the lower quality product will have broader appeal. Athena has an advantage in producing high quality systems. Zeus expects a large market share because of its name recognition, and can do a superior job of merchandizing a lower quality product. Taking all these things into account, Zeus analysts have estimated market shares under various contingencies. The results are shown in Figure 8.1. Payoffs are in millions of dollars to Zeus and Athena, respectively. Since these estimates are based on facts

†My thanks to Eric Lander and Howard Raiffa of the Harvard Business School.

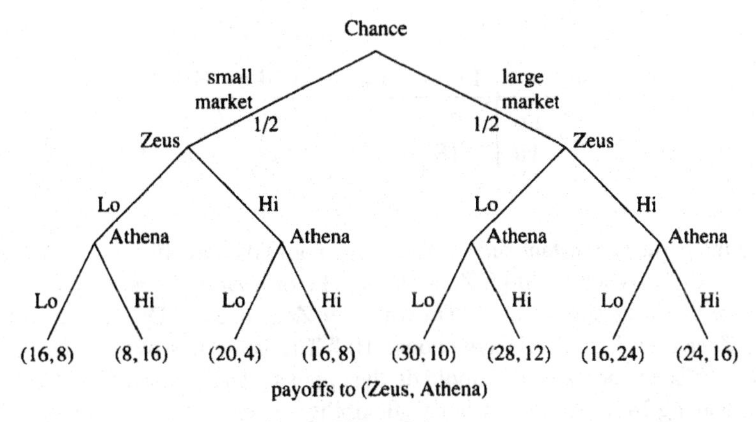

Figure 8.1

known throughout the industry, Zeus has every reason to believe that Athena's analysis of the situation will be very similar.

In Figure 8.1, I have not drawn the information sets. In fact, we will consider how this game should be played under a variety of information conditions. First, suppose that the companies make their production choices in secret, unknown to each other. Of course, neither one knows CHANCE's decision. Then Zeus' two choice nodes are in one information set, and all four of Athena's nodes are in one information set. Each company has only two strategies: produce a high quality system (Hi) or produce a lower quality system (Lo). The game matrix is

		Athena	
		Lo	Hi
Zeus	Lo	(23,9)	(18,14)
	Hi	(18,14)	(20,12)

Game 8.1

The payoffs are expected values. For example, the Lo-Lo payoffs are calculated by $(\frac{1}{2})(16,8) + (\frac{1}{2})(30,10) = (23,9)$. Since the game is constant sum, we can calculate optimal strategies, which are $\frac{2}{7}$Lo, $\frac{5}{7}$Hi for both players. The value of the game is $19\frac{3}{7}$ to Zeus, and hence $12\frac{4}{7}$ to Athena. (It is worthwhile giving some thought to what mixed strategies might mean in this kind of situation.)

Now suppose that Zeus, as industry leader, must make its production decision first, and Athena, as a smaller and more flexible company, can wait to find out Zeus' decision before it makes its own. The effect on the game tree would be that Athena now has two information sets, one containing its first and third nodes (where Zeus has chosen Lo), the other containing its second and fourth nodes (where Zeus has chosen Hi). Since it can make two choices in each of two information sets, it now has four possible strategies. The game is

		Athena			
		Lo/Lo	Lo/Hi	Hi/Lo	Hi/Hi
Zeus	Lo	23	23	18	18
	Hi	18	20	18	20

Game 8.2

Since the game is constant sum, I have shown only the payoff to Zeus. Athena's strategies specify what to do if Zeus chooses Lo or if Zeus chooses Hi. In words, they are "Always choose Lo," "Do whatever Zeus does," "Do the opposite of what Zeus does," and "Always choose Hi." The game has two saddle points. Whatever Zeus does, Athena should do the opposite, and the payoff will be 18 to Zeus, leaving 14 for Athena. Athena's flexibility in being able to wait while Zeus moves first has cost Zeus \$1.43 million. In Chapter 14 we will study the effect of having one player move first in a more general, non-constant-sum context.

Now suppose that Zeus must still move first, but before it makes its production decision, it will conduct an extensive (and expensive) market survey to determine whether the market will be small or large. Athena will not know the result of this survey, but it will know that Zeus has done the survey, since such things are hard to hide. The effect on the game tree would be that Zeus' two nodes are now in separate information sets. Zeus now has four strategies, and the game is

		Athena			
		Lo/Lo	Lo/Hi	Hi/Lo	Hi/Hi
Zeus	Lo/Lo	23	23	18	18
	Lo/Hi	16	20	12	16
	Hi/Lo	25	23	24	22
	Hi/Hi	18	20	18	20

Game 8.3

Athena's strategies are still contingent on whether Zeus chooses Lo or Hi; Zeus' strategies are contingent on whether CHANCE chooses small or large. In this game, Zeus' strategy Hi/Lo, which says "Produce a high quality system if you find the market will be small, otherwise produce a lower quality system," is dominant. Athena should produce a high quality system regardless of what Zeus does, and the expected payoffs will be 22 to Zeus, 10 to Athena.

There are interesting things to note about this solution. First, Athena's optimal strategy has changed from doing the opposite of what Zeus does, to always choosing Hi. Why is this, since Athena has no new information except that Zeus has done a survey? Well, if Zeus chooses Hi, Athena would be better off choosing Lo only if the market will be large. But knowing about the survey, Athena can see that if Zeus chooses Hi, it could only be because the survey says the market will be small. In this case Athena will be better off choosing Hi also. The reasoning is subtle!

Second, conducting the survey and being able to base its production decision on the results increases Zeus' expected profit from \$18 million to \$22 million. If the survey costs less than \$4 million, it is a worthwhile investment.

If Zeus' survey is beneficial to Zeus, how about Athena conducting its own survey? The effect on the game tree would be to put each of Athena's nodes in its own information set. Athena could make its production choice contingent on both Zeus' choice and CHANCE's choice, and it would have 16 possible strategies. It might be good practice to write out the resulting 4×16 game, but it is easier to work directly with the game tree of Figure 8.1. The truncation technique gives the solution of the left half-tree as $(16, 8)$, and the solution of the right half-tree as $(28, 12)$. Since each of these two games will be played with probability $\frac{1}{2}$, the expected payoffs are $(\frac{1}{2})(16, 8) + (\frac{1}{2})(28, 12) = (22, 10)$. Unfortunately, this is no better for Athena than the solution to Game 8.3. Athena should not finance its own consumer survey; it can do just as well by knowing that Zeus will do a survey, and thinking game-theoretically.

Let us end by considering one interesting possibility which cannot be represented by just varying the information sets in Figure 8.1. Suppose that Zeus could conduct its market survey *without Athena's knowledge*. The effect would be that Athena would have incorrect information about which game is being played. Athena would think it is playing Game 8.2, whereas Zeus would know that the actual game is Game 8.3. Athena would play its optimal strategy in Game 8.2, which is Hi/Lo. Knowing this, Zeus would play Hi/Lo and win 24 in Game 8.3, instead of the value of 22 for that game. In other words, if Zeus could keep secret not just the results, but the existence of its survey, it would be worth an extra \$2 million. Having information in a game can be valuable. Knowing better than your opponent who has what information—in other words what game you are playing—can also be valuable.

The analysis of the role of information in competitive decision-making is an important topic in microeconomics. For additional case studies, see [Kreps, 1990a], [Kreps, 1990b], and [Rasmussen. 1989].

Exercises for Chapter 8

Consider three other variations on the theme of Zeus and Athena.

1. Suppose that the companies move simultaneously, but that before they do, Zeus conducts a market survey. Athena knows the existence of the survey, but not its results.
   a) What are the information sets for this game?
   b) Write and solve the resulting 4×2 matrix game.
   c) What effect would it have if Athena did not know of the existence of the survey?

2. Suppose that Athena must move first (because its smaller size makes it less flexible?), but that neither side knows CHANCE's move.
   a) This game requires a different tree. Draw it, showing the information sets.
   b) Write and solve the resulting 4×2 matrix game.
   c) How much would it be worth to Zeus to have Athena move first, rather than having Zeus move first?

3. Suppose that Zeus moves first, but doesn't do a market survey, and Athena does a market survey. Zeus knows the existence of the survey but not its results.
   a) What are the information sets for this game?
   b) How many strategies do Athena and Zeus have?
   c) Solve the game. (If you can do it directly from the game tree it will save you time.)
   d) What effect would it have if Zeus did not know the existence of the survey?

9. Utility Theory

In the discussion so far, we have mostly assumed that the numerical payoffs in our game matrices or game trees are given. We have not paid much attention to where the numbers come from or exactly what they mean. It is time to consider more thoroughly the process of assigning numbers to outcomes, for the applicability of game theory to real situations rests on the assumption that this can be done in a reasonable way. Von Neumann and Morgenstern were very conscious of this dependence, and they began the *Theory of Games and Economic Behavior* by laying the groundwork for modern *utility theory*, the science of assigning numbers to outcomes in a way which reflects an actor's preferences. We need to understand the fundamental ideas of this theory as it affects game theory.

To begin, we should ask what properties the numbers in our matrices must have for the prescriptions of game theory to make practical sense. There are two cases. First, suppose that the game we are considering has a saddle point, a numerical entry which is the largest entry in its column and the smallest entry in its row. This simply means that Rose prefers this outcome to any other entry in its column, and prefers any other entry in its row to it. Hence all that is necessary for game theory to make a sensible recommendation is that the numbers represent the *order* in which Rose prefers the outcomes (and hence the reverse of the order in which Colin prefers them). For example, suppose a 3 × 2 game has outcomes denoted by letters as follows:

		Colin	
		A	B
Rose	A	u	v
	B	w	x
	C	y	z

Suppose we ask Rose to rank these outcomes in order from most preferred to least preferred. Ties are allowed. It is possible that Rose might not be able to do this. For instance, she might say that she prefers u to v, and v to w, but that she also prefers w to u. In such a case of *intransitive preferences* no ordering would be possible. Or she might say, "I simply cannot compare x to y. It's not that I am indifferent between them; it's just that they are so different that I cannot say which I prefer." On the other hand, Rose might be able to give us an ordering, say u, w, x, z, y, v.

Now we go to Colin and ask him to rank the outcomes. For zero-sum theory to be applicable, Colin must be able to do this and his ordering must be the reverse of Rose's, hence v, y, z, x, w, u. If he does this, we can assign numbers to the outcomes by giving a largest number to u, a second largest to w, and so on. For example,

		Colin	
		A	B
Rose	A	6	1
	B	5	4
	C	2	3

The game has a saddle point at BB. Notice that the saddle point would exist and be the same if we varied the numbers by any *order preserving transformation.* Any order preserving transformation would also preserve dominance (e.g., that Rose B dominates Rose C) and higher order dominance.

A scale on which higher numbers represent more preferred outcomes in such a way that only the order of the numbers matters, not their absolute or relative magnitude, is called an *ordinal scale*. Numbers determined from preferences in this way are called *ordinal utilities*. Ordinal utilities are adequate for determining dominance and saddle points.

The second case is when the game we are considering does not have a saddle point and the solution involves mixed strategies. For example, recall from Exercise 1 of Chapter 3 that in the 2×2 game

		Colin		
		A	B	Oddment
Rose	A	a	b	$d-c$
	B	c	d	$a-b$

where a, b, c, and d are numbers with $a > b$ and $d > c$, Rose's optimal strategy is the mixed strategy

$$\frac{d-c}{(d-c)+(a-b)}\text{A}, \quad \frac{a-b}{(d-c)+(a-b)}\text{B}.$$

For this to make sense, the numbers must have been assigned in such a way that the *ratio of differences* $(d-c):(a-b)$ is meaningful. A scale on which not only the order of numbers, but also the ratios of differences of the numbers is meaningful is called an *interval scale*. Numbers reflecting preferences on an interval scale are called *cardinal utilities*. For a mixed strategy game solution to be meaningful, the numbers in the game matrix must be cardinal utilities.

Suppose we are considering the game with possible outcomes u, v, w, and x, and Rose has given us her preference ordering u, x, w, v. To determine Rose's utilities on a cardinal scale, we need to assign numbers to outcomes so that the ratios of differences between the numbers reflect something about Rose's preferences. Von Neumann and Morgenstern noted that we could obtain the relevant information by asking Rose questions about *lotteries*. Start by assigning numbers to v and u arbitrarily, except of course that u gets a higher number than v. For example, assign 0 to v, and 100 to u. Now ask Rose, "Which would you prefer: x for certain, or a lottery which would give you v with probability $\frac{1}{2}$ and u with probability $\frac{1}{2}$?" We will denote such a lottery by $\frac{1}{2}v, \frac{1}{2}u$. If Rose says she would prefer x to the lottery, that means that x ranks higher than the midpoint between v and u, so we must assign x a number higher than 50:

v		x in here	u
0	50		100

Now ask, "Would you prefer x for certain, or the lottery $\frac{1}{4}v, \frac{3}{4}u$?" Suppose that this time Rose prefers the lottery. Then we know that x is ranked below $\frac{3}{4}$ of the way from v to u, i.e., below 75:

v		x in here		u
0	50		75	100

By continuing to change the lottery odds, we can eventually find a lottery so that Rose is indifferent between x and that lottery. For example, Rose might be indifferent between x and $\frac{4}{10}v, \frac{6}{10}u$. Then we would assign x the number 60, $\frac{6}{10}$ of the way from v to u. Now repeat the process for w. Suppose we end up assigning w the number 20. We then have the outcomes spaced along the number line:

v	w		x		u
0	20	40	60	80	100

and these assignments, if they are consistent, give us a representation of Rose's cardinal utilities for the outcomes. Here consistency means that Rose's choices among lotteries should be as predicted by this spacing. For instance, if we ask Rose to choose between w and $\frac{1}{2}v, \frac{1}{2}x$, Rose would choose the lottery. She would be indifferent between w and the lottery $\frac{2}{3}v, \frac{1}{3}x$. The exact consistency conditions needed for preferences to be described by cardinal utilities are given in [Herstein and Milnor, 1953] or [Davis, 1983].

The idea of using questions about lotteries to assign utilities on a cardinal scale is clever, but it did not come out of thin air. If Colin plays a mixed strategy in a game, then Rose's choice of which strategy to play is essentially a choice between lotteries (see Exercise 2). Information about the players' preferences over lotteries is exactly the right information to be meaningful in a game context.

It is harder to elicit consistent cardinal utilities from a player than to elicit ordinal utilities. Rose might not feel comfortable deciding between lotteries, or she might not be able to do it consistently. In my game theory classes I have asked students questions designed to elicit cardinal utilities for five paintings. About 60% of the students have been able to answer lottery questions consistently enough to describe their preferences by cardinal utilities. Another 30% could give consistent ordinal preferences, but had lottery inconsistencies which precluded assigning cardinal utilities. The remaining 10% had intransitive preferences, so that even ordinal utilities could not be assigned. If we cannot assign cardinal utilities to a player's preferences, we cannot obtain a meaningful mixed-strategy solution to a game.

We have not yet asked what is required of Colin's preferences if the game is to be zero-sum. The necessary condition is that Colin's preferences be the reverse of Rose's for lotteries as well for pure outcomes. If they are, and we chose endpoint values of -100 for u and 0 for v, Colin's cardinal utilities would be

u		x		w	v
-100	-80	-60	-40	-20	0

and the game would be zero-sum. However, it is important to notice that the endpoints for a cardinal utility scale are arbitrary, and we could have chosen them differently without affecting the way the utilities predict lottery choices. For example, the following utility assignments represent the same cardinal preferences for Rose as our original 0–100 assignment:

i)

v	w		x		u
-1	0	1	2	3	4

ii)

v	w		x		u
17	19	21	23	25	27

The transformation from one of these scales to the other can be accomplished by a *positive linear function* $f(x) = ax + b$, where $a > 0$. In this case, the function is $f(x) = 2x + 19$, which takes $-1, 0, 2, 4$ to $17, 19, 23, 27$. Cardinal utilities can be transformed by any positive linear function without changing the information they convey.

The invariance of cardinal utilities under positive linear functions means that some games which do not appear to be zero-sum are in fact equivalent to zero-sum games. The easiest examples are constant-sum games, which become zero-sum if you subtract the constant sum c from one player's utilities. But also consider

		Colin A	Colin B
Rose	A	$(27, -5)$	$(17, 0)$
	B	$(19, -1)$	$(23, -3)$

This is not constant-sum: the payoffs for AA total 22 while those for AB total only 17. However, if we transform Rose's utilities by the positive linear function $g(x) = \frac{1}{2}(x - 17)$ we get

		Colin A	Colin B
Rose	A	$(5, -5)$	$(0, 0)$
	B	$(1, -1)$	$(3, -3)$

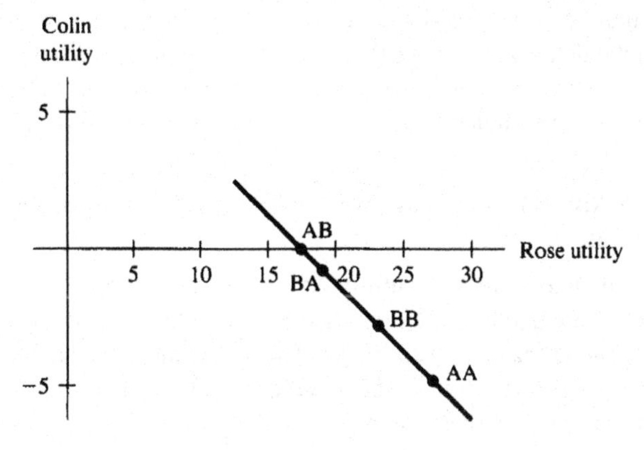

Figure 9.1

which is zero-sum. Hence this game can be solved by zero-sum methods. Figure 9.1 illustrates an easy graphical method for telling if a game is equivalent to a zero-sum game. Plot the payoffs for each outcome in a coordinate plane. If all the points lie along a line of negative slope, then the game is equivalently zero-sum.

The term "utility" has a long history in economic theory, and the word has often been used in questionable ways. In particular, it has not always been used in the same sense as the von Neumann and Morgenstern concept we have discussed here. In the nineteenth century Jeremy Bentham and other "utilitarians" called for economic policies which would maximize the sum of the utilities of all members of society. More recent economists have hypothesized a "rational man" whose goal is to maximize his utility. In our view of utility, Bentham's goal is meaningless, and the idea of rational economic man must be interpreted very carefully. Hence it seems wise to exercise our understanding of von Neumann-Morgenstern utility as a way of organizing a person's consistent responses to questions about lotteries, by considering some possible, and I am afraid common, fallacies about utility. See [Luce and Raiffa, 1957] and [Davis, 1983] for other discussions.

FALLACY #1 (REVERSE CAUSALITY). When a person chooses one alternative over another, or one lottery over another, it is because that person has a higher utility for the chosen alternative.

The problem here is that the order of causality has been reversed. The correct relation is that we assign alternative A higher utility than alternative B *because* our person has indicated that he would choose A over B. This indication may have been direct, as an answer to a question, or indirect, an inference from other choices and the general consistency of choices.

FALLACY #2 (RATIONALITY). In choosing between two alternatives, if a person doesn't choose the one with higher utility, the person is acting irrationally.

It is rather that he is making a choice which is not consistent with the choices from which we determined his utilities. Either his preferences have changed, or his preferences cannot be assigned consistent cardinal utilities. Recall that this was true, in a simple choice situation, for 40% of a group of people I consider highly rational.

FALLACY #3 (ADDING UTILITIES). We can determine socially better alternatives by adding utilities.

For example, if alternative A has utility 20 for Rose and 50 for Colin, while B has utility 100 for Rose and 20 for Colin, then B is socially better than A because its total utility is larger than A's ($120 > 70$). This is Bentham's idea, and it founders on the fact that cardinal utility is only defined up to scale changes by positive linear functions. Such a scale change on either A's or B's utility could easily reverse the outcome in the example.

FALLACY #4 (INTERPERSONAL COMPARISON OF UTILITY). If one person's utility for an outcome is higher than another's, the first person values that outcome more than the second person does.

This statement is nonsense because utilities are arbitrary up to positive linear functions. In fact, we have no operational way to compare two different people's utilities. Utilities are defined only for an individual, and refer to how that individual makes choices among alternatives. Although in everyday life we sometimes make statements like "$100 means more to me than it does to you," it is not at all clear what such statements mean. For instance, if that statement is taken to mean "I would cut off my right hand for $100 and you would not cut off yours," it may just mean that I value my right hand "less" than you value yours. In game theory, we will try to avoid interpersonal comparison of utility as much as we can, because of the risk of meaninglessness in such comparisons. In particular, notice that the definition of a game being zero-sum does not rely on adding or comparing utilities, but only on Rose and Colin making opposite choices between lotteries.

The general viewpoint to remember is that von Neumann-Morgenstern utility is not a quantity with an independent existence. It is simply a convenient numerical way of organizing information about a person's preferences, when those preferences satisfy certain consistency conditions. It puts preference information into exactly the appropriate form for game-theoretic analysis.

Exercises for Chapter 9

1. If Rose's utilities for v, w, x, and u are 0, 20, 60, and 100, respectively, say which alternative Rose would prefer in each of the pairs:
   a) x vs. $\frac{3}{4}w, \frac{1}{4}u$
   b) x vs. $\frac{1}{2}w, \frac{1}{2}u$
   c) $\frac{1}{2}w, \frac{1}{2}x$ vs. $\frac{1}{2}v, \frac{1}{2}u$
   d) $\frac{3}{7}w, \frac{4}{7}x$ vs. $\frac{3}{7}u, \frac{4}{7}v$

2. Suppose that Rose and Colin play the game

		Colin A	Colin B
Rose	A	u	v
	B	w	x

where Rose's utilities are as in Exercise 1 and Colin's utilities are the opposite.

a) What is Colin's optimal strategy in this game?

b) Show that if Colin plays optimally, Rose's expected outcomes for Rose A and Rose B are lotteries, and Rose is indifferent between these two lotteries.

3. In the game

		Colin A	Colin B
Rose	A	r	s
	B	t	u

we have the following information about the players' preferences among the outcomes:

Rose prefers t to s to r to u.

Rose is indifferent between s and the lottery $\frac{2}{3}t, \frac{1}{3}r$.

Rose is indifferent between r and the lottery $\frac{1}{2}s, \frac{1}{2}u$.

Rose's other preferences are consistent with these.

Colin's preferences are the opposite of Rose's.

a) How should the players play this game?

b) If Rose were offered a choice between having s for certain or playing this game, which would she choose? How do you know?

4. Are the following games equivalent to zero-sum games? If so, find a positive linear transformation of one of the player's utilities which shows the equivalence. If not, say how you know.

a)

		Colin A	Colin B
Rose	A	(0, 10)	(1, −10)
	B	(3, −50)	(−1, 30)

b)

		Colin A	Colin B
Rose	A	(−2, 3)	(2, 0)
	B	(3, −2)	(0, 2)

5. Given these utility scales for Rose and Colin

Rose:	p	q	r	s
	0	40	50	100

Colin:	q	p	s	r
	0	20	60	100

which of the following statements are meaningful? Explain!

a) Rose values s more than Colin does.

b) Rose prefers s to r more strongly than Colin prefers r to s.

c) Rose prefers s to q more strongly than Rose prefers q to p.

d) For the society of Rose and Colin, s is a better social choice than r.

e) For the society of Rose and Colin, s is a better social choice than p.

10. Games Against Nature

In our discussion of Jamaican fishing in Chapter 4, we noted that when one player in a two-person zero-sum game is not a reasoning entity capable of the forethought and adaptive play which game theory assumes of players, the minimax solution concept may not apply. On the other hand, we argued that it may still be applicable, depending on the goals of the rational player in the game. In this chapter we will examine in greater detail possible ways of playing a game against Nature, an unreasoning entity whose strategic choice affects your payoff, but which has no awareness of, or interest in, the outcome of the game. In doing so, we will be making a brief foray into the domain of modern statistical decision theory. Our guide for this short visit will be a masterful paper by John Milnor [1954].

Suppose you play the following game against nature:

		Nature			
		A	B	C	D
You	A	2	2	0	1
	B	1	1	1	1
	C	0	4	0	0
	D	1	3	0	0

Which strategy should you choose? We first note that there are two possible situations. The first is when you have enough knowledge about Nature to assign probabilities to her choice of strategies, and you are content to maximize your expected payoff. In this case the expected value principle tells you how to proceed: use your probability estimates to calculate expected values for your strategies, and choose the strategy with the highest expected value. For example, if you believe that the probabilities of Nature choosing A, B, C, D are .2, .4, .3, .1, respectively, then you calculate expected values for your strategies as

$$\begin{aligned}
&A: \quad .2(2) + .4(2) + .3(0) + .1(1) = 1.3\\
&B: \quad .2(1) + .4(1) + .3(1) + .1(1) = 1.0\\
&C: \quad .2(0) + .4(4) + .3(0) + .1(0) = 1.6 \leftarrow \text{highest}\\
&D: \quad .2(1) + .4(3) + .3(0) + .1(0) = 1.4
\end{aligned}$$

and you would choose strategy C. If you were not sure about your assignment of probabilities, it would be wise to do some *sensitivity analysis* by varying the probabilities within ranges you consider reasonable and recalculating expectations to see if your choice would change. For example, if you thought that the probabilities might instead be .3, .3, .3, .1 you would recalculate expectations, getting 1.3, 1.0, 1.2, 1.2, and recognize that strategy A might be an alternate best choice.

The other basic situation is when your uncertainty is so great that you feel you cannot assign probabilities to Nature's strategies in a meaningful way. You simply have no idea how likely or unlikely various states of nature are. This was true in the short run for the Jamaican fishermen. There are a number of suggestions in the literature for how to proceed in this situation. Following Milnor, we will consider four of them.

The oldest suggestion comes from Pierre-Simon de Laplace in his classic *Théorie Analytique des Probabilités* in 1812. Laplace argued that if you have no reason to assume one event is more likely than another, you can do no better than to assume that they are equally likely. He made this "principle of insufficient reason" the basis of his theory of probability. In our context, Laplace's advice is, in the absence of evidence to the contrary, to assume that the four states of nature are equally likely and use the expected value principle. In other words, his advice is

LAPLACE (1812). Choose the row with the highest average entry. Equivalently, choose the row with the *highest row sum.*

Since the row sums in our matrix are 5, 4, 4, 4, Laplace would advise us to choose strategy A.

In the Jamaican fishing example, we thought that the villagers might want to be cautious, considering the worst that could happen to them and making that worst as good as possible. In other words, they should play the maximin strategy. This recommendation is associated with the name of Abraham Wald, one of the founders of modern decision theory.

WALD [1950]. Write down the minimum entry in each row. Choose the row with the largest minimum.

Since the row minima in our matrix are 0, 1, 0, 0, Wald's advice would be to choose strategy B. Notice that Wald's method chooses the saddle point strategy in a game (such as our example) which has a saddle point. If the game does not have a saddle point, it chooses a pure strategy (the maximin strategy) which will *not* be the mixed strategy game theoretic solution. This is very reasonable when the choice is to be made just once—after all, Nature is not going to take advantage of knowing your choice. However, if the game is going to be played many times or by many related people as in the Jamaican fishing case, choosing the optimal game theoretic mixed strategy might be the appropriate way to implement Wald's philosophy.

The maximin principle assumes that the worst will happen; it would seem to be a principle for pessimists. The corresponding principle for optimists would be the *maximax principle* of assuming that the best will happen, and making this best as good as possible. Milnor credits economist Leonid Hurwicz with the idea of mixing these two approaches.

HURWICZ. Choose a "coefficient of optimism" α between 0 and 1. For each row, compute

$$\alpha(\text{row maximum}) + (1 - \alpha)(\text{row minimum}).$$

Choose the row for which this weighted average is the highest.

For example, suppose we choose $\alpha = \frac{3}{4}$. We then compute for our matrix

A: $(\frac{3}{4})(2) + (\frac{1}{4})(0) = 1.5$

B: $(\frac{3}{4})(1) + (\frac{1}{4})(1) = 1.0$

C: $(\frac{3}{4})(4) + (\frac{1}{4})(0) = 3.0$ ← highest

D: $(\frac{3}{4})(3) + (\frac{1}{4})(0) = 2.25$

Hence we should choose strategy C.

A fourth approach to decision under uncertainty, due to L. J. Savage [1954], is based on another psychological observation about how people make decisions. Suppose I decide on strategy C, and Nature chooses strategy A. My payoff is 0. I feel *regret*, because if I had only known what Nature was going to do, I could have chosen my strategy A and gotten 2 instead of 0. Savage proposed that a person might choose so as to *minimize the maximum regret* he might feel after Nature's choice becomes known. Specifically, start from the original matrix and write down the *regret matrix*, in which each entry is the difference between the corresponding entry in the original matrix and the largest entry in its column (which is the best you could have done, had you known Nature's choice). In our example the regret matrix is:

		Nature				
		A	B	C	D	Row maximum
You	A	0	2	1	0	2
	B	1	3	0	0	3
	C	2	0	1	1	2
	D	1	1	1	1	1← smallest

Regret matrix

SAVAGE [1954]. For the regret matrix, write down the largest entry in each row. Choose the row for which this largest entry is smallest.

Savage would advise us to choose strategy D, which guarantees that we will never do more than 1 worse than the best we could have done if we had known Nature's choice. This minimax regret method is worth thinking about. Have you ever made a decision for something like this reason?

At this point we have four possible methods for choosing a strategy in games against nature, and an example for which, as I expect you have noticed, these four methods produce four different recommendations! What should we do? One way of handling this problem is the method of introspection: look inside yourself and see which method best matches your personality. Would you rather have a high average payoff, the certainty that the worst won't be too bad, a chance for the very best, or the guarantee of not too much regret?

A more sophisticated way of trying to choose among the methods is to write down a set of criteria which a good method of playing games against nature should satisfy. Mathematicians call such criteria *axioms*. Then see which of the methods we have considered satisfy which axioms, and make your judgement among methods on that basis. In other words, use the method which satisfies the axioms which you consider most important. The choice of methods again comes down to your preferences, as of course it must, but the hope is that the intervening process of studying axioms will make the behavior of the different methods clearer to you than it would have been without such a study. This *axiomatic method* is something we will return to several times later in this book.

As an example of the axiomatic method, let us consider several axioms due to Milnor [1954]. What properties should a good method of playing games against nature have? Here are three pretty obvious ones:

AXIOM 1: SYMMETRY. Rearranging the rows or columns should not affect which strategy is recommended as best.

AXIOM 2: STRONG DOMINATION. If every entry in row X is larger than the corresponding entry in row Y, a method should not recommend strategy Y.

AXIOM 3: LINEARITY. The recommended strategy should not change if all entries in the matrix are multiplied by a positive constant, or if a constant is added to all entries.

Axiom 3 just says that the entries in the matrix are to be thought of as utilities. You can easily check that all four of the decision methods we have considered satisfy all of these three axioms. Here are some axioms which differentiate among the methods:

AXIOM 4: COLUMN DUPLICATION. The recommended strategy should not change if we add to the matrix a new column which is a duplicate of a column already in the matrix.

After all, a state of nature does not become more likely just because we choose to list it twice. The Wald, Hurwicz, and Savage methods all satisfy Axiom 4, but the Laplace method violates it. For example, if we duplicated column B in our matrix, the Laplace method would change its recommendation from strategy A to strategy C.

AXIOM 5: BONUS INVARIANCE. The recommended strategy should not change if a constant is added to every entry in some column.

This condition says that if Nature decides to give you a bonus, or exact a penalty, which depends on her choice but not on yours, that should not affect what you do. Does that seem reasonable? The Laplace method satisfies Axiom 5, since all of the row totals would be increased by the same amount. The Savage method satisfies the axiom because the regret matrix is not changed. However, the Wald and Hurwicz methods violate Axiom 5. See Exercise 2.

AXIOM 6: ROW ADJUNCTION. Suppose a method chooses row X as the best strategy to follow in a game against nature, and then a new row Z is added to the matrix. The method should then choose either X or Z, but not some other row.

The idea here is that if X was better than any other previously available strategy, the discovery of a new strategy Z should not affect that. Laplace, Wald and Hurwicz all satisfy Axiom 6, but the Savage method violates it, as you are asked to show in Exercise 3.

If we believe that all of these axioms are necessary conditions for a good decision method, we must conclude that none of the four methods we have considered is good, since each violates one of the axioms. We are left with two choices. If we are creative, we might try to come up with a better method which does satisfy all six axioms, and any other axioms we might come up with. The warning here is that there may be no such method—Milnor gives ten axioms and proves that *no* method can satisfy all of them.

Alternatively, we could agree that for certain kinds of games against nature we are willing to sacrifice one of the axioms, and hence find an acceptable choice method. The axioms which the methods violate can be thought of as cautions: if some axiom is important in a particular kind of game against nature, don't play that game using a method which violates that axiom. For example, the Laplace method should not be used when there is ambiguity in delineating possible states of nature, because how you delineate states of nature can affect the strategy you choose. The Wald and Hurwicz methods should not be used if there are uncertainties in your estimates of payoffs which depend upon the state of nature but not upon your strategy. The Savage method should not be used if you are not sure what strategies are really available to you, for adding or subtracting even unattractive strategies can affect your choice among attractive strategies.

These kinds of conclusions, which come out of the axiomatic analysis, would not have been obvious from a more purely psychological characterization of the methods. There is power in the axiomatic approach.

Exercises for Chapter 10

1. Find what each decision method (Laplace, Wald, Hurwicz with a coefficient of optimism suitable for you, and Savage) would tell a company manager to do in the following

decision situation. The manager has no information about what the economy will be like three years from now when the payoff will come.

		Economy			
		Way up	Slightly up	Slightly down	Way down
Manager	Hold steady	3	2	2	0
	Expand slightly	4	2	0	0
	Expand greatly	6	2	0	−2
	Diversify	1	1	2	2

(profit to company in $million)

2. a) Show that the Wald method violates the Bonus invariance axiom by considering the effect of adding 2 to every entry in column C, and 1 to every entry in Column D, in the example matrix of this chapter.
   b) Show that the Hurwicz method violates Bonus invariance by considering the effect of adding a suitable bonus to some column of the matrix (use $\alpha = \frac{3}{4}$).

3. Show that the Savage method violates the Row adjunction axiom by considering the effect of adjoining Row E: 0 0 0 3 to the example matrix.

4. Show that if some strategy X is chosen by the maximax method (Hurwicz with $\alpha = 1$), then
   a) giving certain column bonuses will make X be chosen by the Wald method;
   b) duplicating certain columns will make X be chosen by the Laplace method.

Part II

Two-Person Non-Zero-Sum Games

11. Nash Equilibria and Non-Cooperative Solutions

If a two-person game is not zero-sum, we must write both players' payoffs to describe the game. If we have a game in which the payoffs to the players do not add to zero, recall from Chapter 9 that the game still might be equivalent to a zero-sum game, in the sense that it could be made zero-sum by a change of utility scales. In such a game, the interests of the two players are strictly opposed, and we can analyze it by zero-sum methods. In general, however, the interests of players in a non-zero-sum game are not strictly opposed, and not strictly coincident. The game will combine competitive aspects with some opportunities for cooperation.

Cooperation may require communication between the players, and hence the analysis of non-zero-sum games depends on what kind of assumptions we make about how the players can communicate with each other. We will make three different levels of assumptions. In this chapter we will assume that no communication between players is possible. Play will be exactly as in the zero-sum case: the players choose their strategies simultaneously, and their choice is not known to the other player. In Chapter 14 we will allow the players to talk to each other before they choose strategies, opening the way for such strategic gambits as commitments, threats and promises. Finally, in Chapter 16 we will investigate how players might agree on cooperative solutions which are in some sense "fair" to both players.

For the non-cooperative analysis of non-zero-sum games, the logical place to begin is with our work on zero-sum games. To what extent can ideas from zero-sum games carry over to non-zero-sum games? As a first example, consider

		Colin A		Colin B
Rose	A	(2, 3)	←	(3, 2)
		↑		↑
	B	(1, 0)	→	(0, 1)

payoffs to (Rose, Colin)

Game 11.1

If we look at this game from Rose's point of view, we notice that no matter which strategy Colin chooses, Rose will be better off choosing Rose A than Rose B. Rose A *dominates* Rose B, and we would expect Rose to choose Rose A. Knowing this, Colin should of course choose Colin A rather than Colin B, and we would predict (2, 3) as the outcome. Although the payoffs look symmetric, the arrangement of the payoffs favors Colin.

This example shows that the Dominance Principle from zero-sum theory still applies in the non-zero-sum context. So does the idea of the movement diagram, shown above, and the idea of an equilibrium outcome. AA is identified as an equilibrium outcome because it has only incoming arrows.

Equilibrium outcomes in non-zero-sum games correspond to saddle points in zero-sum games, and just as there are zero-sum games without saddle points, there are non-zero-sum games without pure-strategy equilibria:

		Colin A		Colin B
Rose	A	(2,4)	←	(1,0)
		↓		↑
	B	(3,1)	→	(0,4)

Game 11.2: A game with no pure-strategy equilibrium

The movement diagram is shown in the matrix, and it is clear that there is no pure-strategy equilibrium. In the zero-sum case we solved this problem by considering mixed strategies. Hence we should ask here: are there mixed strategies for Rose and Colin such that if they both play these strategies, neither could gain by switching to some other strategy? There are indeed such strategies. Suppose we consider *Colin's game*, which is the zero-sum game (with payoffs to Colin) obtained by considering only Colin's payoffs in Game 11.2. Rose's optimal mixed strategy in this game is $(\frac{3}{7}A, \frac{4}{7}B)$, and this strategy has the property that it will give Colin an expected payoff of $\frac{16}{7}$ regardless of which strategy Colin plays. Let's call it Rose's *equalizing strategy*. Similarly, Colin's equalizing strategy of $(\frac{1}{2}A, \frac{1}{2}B)$, obtained by looking at Rose's game, will give Rose an expected payoff of $\frac{3}{2}$ regardless of what Rose does.† Hence if both players adopt these equalizing strategies, neither player can gain by deviating. We have found an equilibrium in mixed strategies. John Nash [1950a] proved that every two-person game has at least one equilibrium in either pure strategies or mixed strategies. We call equilibria in non-zero-sum games *Nash equilibria* in his honor.

If we could use the idea of a Nash equilibrium as our solution to a non-zero-sum game, the theory of non-zero-sum games would be not much harder than the zero-sum theory. Unfortunately, there are games where the Nash equilibria

---

†It is important to note that Colin's equalizing strategy here is *not* Colin's optimal strategy in Rose's game. Rose's game has a saddle point at AB. Hence if we apply the method of oddments to it, getting $(\frac{1}{2}A, \frac{1}{2}B)$ for Colin, the result is not the optimal strategy. It *is*, though, a strategy which equalizes Rose's expected payoffs for Rose's two strategies, and that is what we are after here.

have unattractive and problematical properties. We can see one such property in the Nash equilibrium of Game 11.2. We found that equilibrium by having each player play in the *other* player's game. Each player ignores his or her own payoffs, and plays only to equalize the other player's payoffs. It is no wonder that the resulting payoffs of $(1.50, 2.29)$ seem low for both players. Surely the players could do better by paying attention to their own payoffs. We will return to this point, but first let us consider some other examples.

		Colin A		Colin B
Rose	A	(1, 1)	→	(2, 5)
		↓		↑
	B	(5, 2)	←	(−1, −1)

Game 11.3: A game with two non-equivalent and non-interchangeable equilibria.

In this game there are two different pure-strategy equilibria, AB and BA. Recall that a zero-sum game could have multiple saddle points, but those saddle points were always *equivalent* and *interchangeable*: all saddle points had the same value, and if both players played a saddle point strategy, the resulting outcome was always a saddle point. Game 11.3 shows that neither of these properties hold for non-zero-sum games. In fact, the equilibrium at BA is better for Rose, the equilibrium at AB is better for Colin, and if both players try for their favorite equilibrium, they will end up at BB, which is not an equilibrium and is the worst outcome for both players. If a game has several non-equivalent and non-interchangeable Nash equilibria, it may not be clear which equilibrium the players should try for.

Finally, consider

		Colin A		Colin B
Rose	A	(3, 3)	→	(−1, 5)
		↓		↓
	B	(5, −1)	→	(0, 0)

Game 11.4: A game with a unique equilibrium, which is not Pareto optimal.

This game has a unique Nash equilibrium at BB. In fact, Rose B dominates Rose A, and Colin B dominates Colin A, so this is an equilibrium of the strongest possible type. However, it does not seem to be a very happy outcome, since both Rose and Colin would be better off at AA, getting a payoff of 3 instead of 0. About 1900 the Italian economist Vilfredo Pareto proposed that we should not accept an economic system if there is another available system which would make everyone better off. If we adapt this idea to two-person games, we get:

DEFINITION. An outcome of a game is *non-Pareto-optimal* (or Pareto inferior or Pareto inefficient) if there is another outcome which would give both players higher payoffs, or would give one player the same payoff but the other player a higher payoff. An outcome is *Pareto optimal* if there is no such other outcome.

Note that the word "optimal" here does not mean "best," just "not obviously inferior to some other outcome." In general a game will have many Pareto optimal outcomes. In fact, in a zero-sum game *every* outcome is Pareto optimal, since every gain to one player means a loss to the other. In Game 11.4, outcomes AA, AB, and BA are all Pareto optimal. Only BB is non-Pareto-optimal, since AA would give both players higher payoffs. Pareto's idea can be formulated as the

PARETO PRINCIPLE. To be acceptable as a solution to a game, an outcome should be Pareto optimal.

The Pareto Principle is a cogent principle of *group rationality*. Unfortunately, in Game 11.4 it comes into direct conflict with an equally cogent principle of *individual rationality*, the Dominance Principle. Because it embodies this conflict, Game 11.4 is a central example in non-zero-sum theory, and we will return to it in Chapter 12.

It is easy to see which outcomes of a game are Pareto optimal if we plot the outcomes in a coordinate plane, where the horizontal coordinate is Rose's payoff and the vertical coordinate is Colin's payoff. After we have plotted the points representing the pure-strategy outcomes, mixed strategy outcomes are represented by points in the convex polygon enclosing the pure-strategy points. For example, the outcome $\frac{1}{2}$AA + $\frac{1}{2}$BB appears as the midpoint of the line segment joining AA to BB. This polygon is called the *payoff polygon* for the game. Figure 11.1 shows the payoff polygons for Games 11.1 through 11.4. The Pareto optimal outcomes are exactly those which lie on the "northeast" boundary of the payoff polygon. In Figure 11.1 the Pareto optimal outcomes are dashed. Notice that the set of Pareto optimal outcomes for a game can be a line segment, several line segments, or even just a single point.

Recall that I suggested that the mixed-strategy Nash equilibrium for Game 11.2 was not a very good outcome. If we plot it in the payoff polygon, as shown in Figure 11.2b, we can see why it isn't: it is not Pareto optimal. The pure outcome AA, and many mixtures of AA and BA, would be better for both players.

What we have seen so far is that there are serious problems with the idea of an equilibrium point as a solution concept for non-zero-sum games. An equilibrium outcome is certainly desirable because of its stability, and Nash proved that one always exists. However, there may be multiple equilibrium points which are non-equivalent and non-interchangeable, giving rise to coordination problems. Even if there is a unique equilibrium point, it may not be Pareto optimal. Given these problems, perhaps we should try a different idea. Recall that in zero-sum games, equilibrium outcomes arose precisely when both players played their cautious

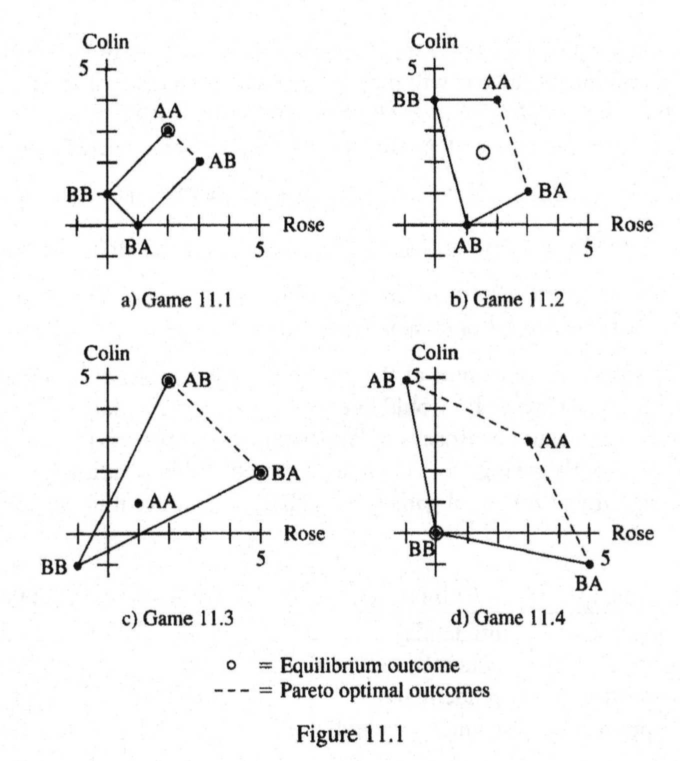

Figure 11.1

minimax strategies of maximizing their payoff in the worst possible case. Let us explore the behavior of minimax strategies in the non-zero-sum context.

Consider Rose's choice in Game 11.2. The worst possible case would be if Colin played to hold Rose's payoff down, and we already know what Rose should do to minimize the damage in that case: she should play the minimax strategy in Rose's game. Rose's game has a saddle point at AB, so Rose should play Rose A, assuring her at least a payoff of 1, the value of Rose's game. We make the following

DEFINITION. In a non-zero-sum game, Rose's optimal strategy in Rose's game is called Rose's *prudential strategy*. The value of Rose's game is called Rose's *security level*.

By playing her prudential strategy, Rose can assure that she will get at least her security level. Of course, the definition for Colin is analogous. In Game 11.2 Colin's prudential strategy is the mixed strategy $(\frac{4}{7}A, \frac{3}{7}B)$ and Colin's security level is $\frac{16}{7}$.

If both players play their prudential strategies in Game 11.2 the outcome will be $\frac{4}{7}AA + \frac{3}{7}AB = (\frac{11}{7}, \frac{16}{7})$. Colin gets exactly his security level; Rose does somewhat better than her security level. If we plot this outcome in the payoff polygon, we see that it is not Pareto optimal. It is also not an equilibrium. In fact, if Colin thinks that Rose will play her prudential strategy of Rose A, Colin should play not $(\frac{4}{7}A, \frac{3}{7}B)$, but pure Colin A. Similarly, if Rose thinks Colin will play his prudential mixed strategy, Rose can calculate her expected payoffs for

$$\text{Rose A: } (\tfrac{4}{7})(2) + (\tfrac{3}{7})(1) = \tfrac{11}{7} \qquad \text{Rose B: } (\tfrac{4}{7})(3) + (\tfrac{3}{7})(0) = \tfrac{12}{7}$$

and should play Rose B. We will give these "best response" strategies a name:

DEFINITION. In a non-zero-sum game, a player's *counter-prudential strategy* is his optimal response to his opponent's prudential strategy.

Table 11.1 shows the outcomes of the possible combinations of prudential and counter-prudential play. Colin would like Rose to play prudentially, so he can respond counter-prudentially. Rose would be happy if both players played counter-prudentially. Clearly the logic is complicated and not stable. Cautious play, which produced stability in zero-sum games, fails badly to produce it in non-zero-sum games.

Rose strategy	Colin strategy	Rose payoff	Colin payoff
prudential	prudential	1.57	2.29
prudential	counter-prudential	2.00	4.00
counter-prudential	prudential	1.71	2.29
counter-prudential	counter-prudential	3.00	1.00

Rose prudential:	A
Colin prudential:	$\frac{4}{7}$A, $\frac{3}{7}$B
Rose counter-prudential:	B
Colin counter-prudential:	A

Table 11.1. Prudential and counter-prudential play in Game 11.2.

I am afraid that the result of our discussion must be that the solution theory for zero-sum games does not carry over to non-zero-sum games, and in fact that there is no cogent general solution concept for non-zero-sum games. We simply cannot give a general prescription for how to play all such games when communication between the players is not allowed. However, it might be useful to single out those games for which the problems we have been discussing do not arise, and for which we can give reasonable advice:

DEFINITION. A two-person game is *solvable in the strict sense* (or *SSS*) if

i) there is at least one equilibrium outcome which is Pareto optimal, and
ii) if there is more than one Pareto optimal equilibrium, all of them are equivalent and interchangeable.

For a game which is solvable in the strict sense, we would prescribe the unique Pareto optimal equilibrium, or the set of all equivalent and interchangeable Pareto optimal equilibria, as the solution. Game 11.1 is SSS, but Games 11.2, 11.3, and 11.4 are not. For a final example, consider Game 11.5, whose movement diagram and payoff polygon are shown in Figure 11.2.

		Colin A	Colin B	Colin C
Rose	A	$(0,-1)$	$(0,2)$	$(2,3)$
	B	$(0,0)$	$(2,1)$	$(1,-1)$
	C	$(2,2)$	$(1,4)$	$(1,-1)$

Game 11.5

There are two equilibria, at BB and AC. However, the payoff polygon shows that BB is not Pareto optimal, so AC is a unique Pareto optimal equilibrium. Hence the game is solvable in the strict sense, and we would prescribe that Rose should play Rose A and Colin should play Colin C. These are not, by the way, Rose and Colin's prudential strategies (Exercise 3).

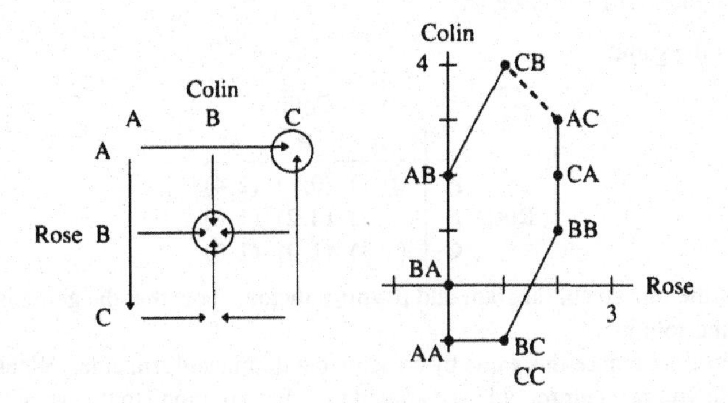

Figure 11.2: Movement diagram and payoff polygon for Game 11.5

Finally, Exercise 4 asks you to consider whether even our quite restrictive definition of which games are "solvable" might still be too broad. In non-zero-sum games, subtlety abounds and easy answers are hard to come by.

Exercises for Chapter 11

1. Find a Nash equilibrium in

		Colin A	Colin B
Rose	A	$(3,2)$	$(2,1)$
	B	$(4,3)$	$(1,4)$

Is it Pareto optimal?

2. For each of the following games, draw the movement diagram and the payoff polygon, and identify all (pure-strategy) equilibria and Pareto optimal outcomes. Determine whether the game is solvable in the strict sense. If so, give the solution. If not, say why not.

a)

		Colin	
		A	B
Rose	A	(2, 2)	(4, 3)
	B	(3, 4)	(1, 1)

b)

		Colin	
		A	B
Rose	A	(2, 2)	(4, 1)
	B	(1, 3)	(3, 4)

c)

		Colin		
		A	B	C
Rose	A	(3, 0)	(5, 2)	(0, 4)
	B	(2, 2)	(1, 1)	(3, 3)
	C	(4, 1)	(4, 0)	(1, 0)

d) The game in c) with the AC payoff changed to (0, 5).

3. For Game 11.5
   a) Verify that Rose's prudential strategy is ($\frac{1}{7}$A, $\frac{2}{7}$B, $\frac{4}{7}$C). What is her security level?
   b) Find Colin's prudential strategy and security level.
   c) What will the expected payoffs be if both players play prudentially? Is this Pareto optimal?
   d) Calculate the counter-prudential strategies for Rose and Colin, and make a table like Table 11.1 for this game.

4. Consider the game

		Colin		
		A	B	C
Rose	A	(0, 1)	(0, 1)	(2, 4)
	B	(5, 1)	(4, 2)	(1, 0)
	C	(4, 3)	(1, 4)	(1, 0)

   a) Draw the movement diagram and payoff polygon, show that the game is SSS, and find the solution.
   b) Suppose we reduce this game by crossing out dominated strategies. What happens?
   c) Would you feel comfortable recommending the "solution" to this game?

12. The Prisoner's Dilemma

In 1950 Melvin Dresher and Merrill Flood at the RAND Corporation devised Game 12.1 to illustrate that a non-zero-sum game could have an equilibrium outcome which is unique, but fails to be Pareto optimal.

		Colin	
		A	B
Rose	A	(0,0)	(−2,1)
	B	(1,−2)	(−1,−1)

Game 12.1: The original Prisoner's Dilemma. A is "don't confess," B is "confess."

Later, when presenting this example at a seminar at Stanford University, Albert W. Tucker told a story to go with the game (see [Straffin, 1980]). The players are two prisoners, arrested for a joint crime, who are being interrogated in separate rooms. The clever district attorney tells each one that

- if one of them confesses and the other does not, the confessor will get a reward (payoff $+1$) and his partner will get a heavy sentence (payoff -2).
- if both confess, each will get a light sentence (payoff -1).

At the same time, each has good reason to believe that

- if neither confesses, both will go free (payoff 0).

In the years since 1950 this game has become known as the Prisoner's Dilemma. It is the most widely studied and used game in social science.

Game 12.1 is ordinally equivalent to Game 11.4, which would thus also be called Prisoner's Dilemma. We have already seen the dilemma. Strategy B is dominant for both players, leading to the unique equilibrium at BB. However, this equilibrium is non-Pareto-optimal, since both players would do better at AA. In terms of the story, each prisoner is better off confessing, no matter what he believes the other prisoner will do. However, if both confess, both are worse off than if neither had confessed. Think about it carefully. The conflict between individual rationality in the form of the Dominance Principle and group rationality in the form of the Pareto Principle is absolute. Individuals rationally pursuing their own best interest end at an outcome which is unfortunate for both of them.

The general form of Prisoner's Dilemma is shown as Game 12.2, with suggestive names for the strategies and payoffs. The first condition guarantees that strategy D dominates strategy C for both players, and that the unique equilibrium

		Colin C	Colin D
Rose	C	(R,R)	(S,T)
	D	(T,S)	(U,U)

C: cooperate
D: defect
R: reward for cooperation
S: sucker payoff
T: temptation payoff
U: uncooperative payoff

Game 12.2: The general form of Prisoner's Dilemma
Conditions: $T > R > U > S$ and $R \geqslant (S + T)/2$
[Rapoport and Chammah, 1970]

at DD is Pareto inferior to CC. The second condition says that the players would be at least as well off always playing CC as alternating between CD and DC, so that CC is Pareto optimal.

The importance of Prisoner's Dilemma lies in the fact that many social phenomena with which we are familiar seem to have Prisoner's Dilemma at their core. Imagine two stores engaged in a price war, deciding whether to cut prices or not. If the other store does not cut prices, I can attract its customers by cutting prices. If it does cut prices, I had better cut prices so as not to lose my customers. If both stores reason this way and cut prices, both make lower profits than if neither had cut prices. For two nations engaged in an arms race, the corresponding strategic choices might be "arm" and "don't arm," and the same reasoning might prevail. I expect that you can think of many other examples.

Because this kind of conflict between individual rationality and group rationality seems so pervasive, a number of attempts have been made to resolve the Prisoner's Dilemma, usually by finding some argument which can justify playing the cooperative strategy C in Game 12.2 (or strategy A in Games 12.1 and 11.4) in defiance of the Dominance Principle. We will consider three of these attempts which are interesting in their own right and throw different kinds of light on the dilemma.

Repeated Play in Theory

Perhaps the most obvious way to try to resolve Prisoner's Dilemma is to note that in many social contexts for which it might be a model, the game is played not just once, but repeatedly. In repeated play, it might be worth cooperating in early plays in hope of arriving at the mutually beneficial outcome CC, rather than face a long string of unprofitable DD's. Unfortunately, this idea falls victim to a logical domino-type argument. Suppose there are to be 100 plays of a Prisoner's Dilemma game, and consider what will happen on the last play. On that play both players will know that strategy D dominates strategy C, and there will be no future plays to make it worthwhile trying to induce mutual cooperation. The outcome on the 100th play will be DD. Once both players realize this, the 99th play becomes in effect the last play, and the same argument applies to it. The plays fall backwards like dominoes, and even the first outcome must be DD. Strict logic prevents cooperation from ever getting started. (Exercise 1 asks you to further investigate this argument by constructing and analyzing the "supergame" of two successive plays of Game 12.1.)

One possible escape from this argument is to observe that, in practice, real players do not adhere to this strict logical course of play. We will return to this observation at the end of this chapter. A second possible escape is to note that in many contexts where a Prisoner's Dilemma-like game is played repeatedly, the players do not know how many games will be played. If they do not know which game will be the last, there is no last domino to start the others falling.

One formal approach to this situation ([Shubik, 1970], [Hill, 1975]) is to assume that after each play of Prisoner's Dilemma, the next play will occur with probability p, where p is some number between 0 and 1. Thus the first play will happen with probability 1, the second with probability p, the third with probability p^2, the nth with probability p^n. Suppose I assume (a big assumption) that my opponent will start by choosing C and will continue to choose C until I choose my first D. After that, we both will choose all D's. If I never choose a D, my payoff will be

$$R + pR + p^2R + p^3R + \cdots = \frac{R}{(1-p)} \tag{1}$$

by the formula for summing a geometric series (Exercise 2). On the other hand, if I decide to choose my first D in the mth game, my payoff will be

$$\begin{aligned} & R + pR + p^2R + \cdots + p^{m-1}R + p^mT + p^{m+1}U + p^{m+2}U + \cdots \\ &= \frac{R(1-p^m)}{(1-p)} + p^mT + \frac{p^{m+1}U}{(1-p)} \qquad \text{by Exercise 2} \\ &= \frac{R - p^mR + (1-p)p^mT + p^{m+1}U}{(1-p)}. \qquad (2) \end{aligned}$$

Hence I should never choose a D if the expression (1) is larger than the expression (2) for all values of m, i.e. if

$$\begin{aligned} \frac{R}{(1-p)} &> \frac{R - p^mR + p^mT - p^{m+1}T + p^{m+1}U}{(1-p)}, \\ 0 &> -p^m(R - T + pT - pU), \\ T - R &< p\,(T - U), \\ p &> \frac{T-R}{T-U}. \qquad (3) \end{aligned}$$

In other words, if the probability of continuing to play is higher than a certain threshold value, it makes sense for both players to cooperate, under the assumption that the other player will also. For Game 12.1 the threshold probability is $(1-0)/(1-(-1)) = \frac{1}{2}$. For Game 11.4 the threshold is $(5-3)/(5-0) = \frac{2}{5}$.

Unfortunately, the beginning assumption I made about my opponent's play makes the above reasoning less than completely convincing, but it does strongly suggest the possibility of mutual cooperation when continuing play is uncertain but likely.

The Metagame Argument

If Prisoner's Dilemma is played just once, is there any way to argue that the players should choose C? Consider Colin. He might want to cooperate, but fear that if he did, Rose would not, and he would get the sucker's payoff. So perhaps he should make his choice *contingent* on what he thinks Rose will do. For example, he could decide to cooperate if he thinks Rose will cooperate, but defect if he thinks Rose will defect. Unfortunately, this hopeful idea runs up against the fact that Colin is better off defecting even if he thinks Rose will cooperate. However, suppose we push this idea one step further. Suppose that Colin will make his choice contingent on what he thinks Rose will do, and Rose will make her choice contingent on what contingent policy she thinks Colin is following. In this complicated scenario, might cooperation arise?

Nigel Howard [1971] formalized this idea as what he called a *metagame*. A clear introductory account of the idea is given in [Rapoport,1967b]. Consider Game 12.1. Let Colin's choice be contingent on Rose's choice, so that Colin has four strategies:

I: Choose A regardless of what he thinks Rose will do,

II: Choose the same alternative he thinks Rose will choose,

III: Choose the opposite alternative of what he thinks Rose will choose,

IV: Choose B regardless of what he thinks Rose will do.

The resulting game is Game 12.3, the *first level metagame* of Game 12.1. Notice that in calculating the payoffs, it is assumed that Colin predicts Rose's choice correctly and acts accordingly.

		Colin			
		I: AA	II: AB	III: BA	IV: BB
Rose	A	$(0,0)$	$(0,0)$	$(-2,1)$	$(-2,1)$
	B	$(1,-2)$	$(-1,-1)$	$(1,-2)$	$(-1,-1)$

Game 12.3: First level metagame of Game 12.1.

In Game 12.3, it is no longer the case that Rose B dominates Rose A. Unfortunately, Colin IV dominates all of Colin's other strategies, and the unique equilibrium is at Rose B-Colin IV, which is still non-cooperative. This corresponds to our first-level argument above, so let us continue to the second level.

Let Rose's choice be contingent on Colin's contingent strategy, so Rose will have 16 strategies. The notation for a Rose strategy in Game 12.4 says what Rose will do for each of the four Colin possibilities. For example, Rose XII, denoted BABB, means that Rose will choose A if she believes Colin is using strategy Colin II, but B if she believes Colin is using Colin I, III, or IV. In calculating the payoffs for Game 12.4, it is assumed that Rose correctly predicts Colin's strategy and acts accordingly, and that Colin correctly predicts the resulting Rose action and acts accordingly!

		Colin			
		I: AA	II: AB	III: BA	IV: BB
Rose	I: AAAA	(0, 0)	(0, 0)	(−2, 1)	(−2, 1)
	II: AAAB	(0, 0)	(0, 0)	(−2, 1)	(−1, −1)
	III: AABA	(0, 0)	(0, 0)	(1, −2)	(−2, 1)
	IV: AABB	(0, 0)	(0, 0)	(1, −2)	(−1, −1)
	V: ABAA	(0, 0)	(−1, −1)	(−2, 1)	(−2, 1)
	VI: ABAB	(0, 0)	(−1, −1)	(−2, 1)	(−1, −1)
	VII: ABBA	(0, 0)	(−1, −1)	(1, −2)	(−2, 1)
	VIII: ABBB	(0, 0)	(−1, −1)	(1, −2)	(−1, −1)
	IX: BAAA	(1, −2)	(0, 0)	(−2, 1)	(−2, 1)
	X: BAAB	(1, −2)	(0, 0)	(−2, 1)	(−1, −1)
	XI: BABA	(1, −2)	(0, 0)	(1, −2)	(−2, 1)
	XII: BABB	(1, −2)	(0, 0)	(1, −2)	(−1, −1)
	XIII: BBAA	(1, −2)	(−1, −1)	(−2, 1)	(−2, 1)
	XIV: BBAB	(1, −2)	(−1, −1)	(−2, 1)	(−1, −1)
	XV: BBBA	(1, −2)	(−1, −1)	(1, −2)	(−2, 1)
	XVI: BBBB	(1, −2)	(−1, −1)	(1, −2)	(−1, −1)

Game 12.4: Second level metagame of Game 12.1.

In Game 12.4 Colin does not have a dominant strategy, but you can check that Rose XII (BABB) dominates all other Rose strategies. Knowing this, Colin should play Colin II, leading to an equilibrium which is cooperative! (Exercise 3 asks you to find the two other equilibria in this game, which do not arise from dominance.) Hence this second level contingent reasoning does lead to cooperation. Rose's best policy is to choose to cooperate if and only if she believes that Colin will match her choice, and knowing this, Colin should match Rose's choice.

I think this reasoning is quite wonderful in its second level complexity: "Cooperate if and only if you believe that your opponent will cooperate if and only if you do." However, I have difficulty with the argument because of the amount of mind reading involved. Recall that to construct the payoff matrices we had to suppose that both players were able to predict each other's choices at an impressive level of intricacy. I just don't believe that is reasonable to assume. In his presentation of Howard's argument, Rapoport wrote:

> In order to be intuitively understood and accepted, this formal solution of the Prisoner's Dilemma paradox still needs to be translated in a social context. When and if that is accomplished, Prisoner's Dilemma will deserve a place in the museum of famous ex-paradoxes. [Rapoport, 1967b]

I take this to mean that we must find some way to carry through the essence of the argument without requiring mind reading, and I simply don't see how to do that. Perhaps you will have better luck. The metagame idea is certainly related to the theory of strategic moves in Chapter 14.

Repeated Play in Practice

In practice, people playing repeated Prisoner's Dilemma do not usually play completely non-cooperatively, even when the number of plays is fixed and strict logic requires all D's. For example, in my most recent game theory class seven pairs of students played 20 rounds of Prisoner's Dilemma, before they learned about the theory of the game. Seven of the 14 players played C at least one-quarter of the time. On the other hand, five of the seven pairs eventually "locked in" and played only DD's in the final rounds, and no pairs locked in on CC. Naive players are willing to experiment with cooperation in early rounds of play, but are rarely able to sustain mutual cooperative play. Rapoport and Chammah [1970] report many results along these lines.

It might be more interesting to know how sophisticated players play repeated Prisoner's Dilemma. If people know the theory of the game and play against other people who they know know it, what will they do? The most ambitious experiment of this kind was run by Robert Axelrod, and the results are reported in [Axelrod, 1984]. Axelrod invited professional game theorists to write computer programs to play iterated Prisoner's Dilemma, and then conducted a tournament in which each program played all the other programs. Fourteen programs were entered, some of considerable complexity. The winner was a four line program submitted by Anatol Rapoport, called TIT FOR TAT. Its instructions were

1. Start by choosing C.
2. Thereafter, in each round choose whatever your opponent chose in the previous round.

In other words, "Do unto your opponent what your opponent has just done unto you." It's a repeated play version of Colin's strategy II in Game 12.3.

Axelrod reported the results of his tournament, including a detailed analysis of the highest scoring programs. He gave an example of a program which would have won the tournament, beating TIT FOR TAT, if it had been entered. He then invited entries for a new tournament. Sixty-two game theorists responded with programs for Round Two, including a number of programs intricately designed to do well against TIT FOR TAT. Anatol Rapoport entered TIT FOR TAT again, unchanged.

The result was unexpected: TIT FOR TAT won again! The clever programs designed to do well against it, did not fare as well when pitted against each other. (Jim Graaskamp and Ken Katzen, two students from my game theory class, wrote a program which came in 6th out of the 63.)

One immediate conclusion is that TIT FOR TAT is a remarkably good and robust strategy for playing iterated Prisoner's Dilemma, at least in an environment where a number of sophisticated players are competing. A more general conclusion comes from Axelrod's analysis of programs which did well. They tended strongly to share four properties with TIT FOR TAT. To do well, a strategy should be

- Nice. It should start by cooperating, and never be the first to defect.
- Retaliatory. It should reliably punish defection by its opponent.
- Forgiving. Having punished defection, it should be willing to try cooperating again.
- Clear. Its pattern of play should be consistent and easy to predict.

It is tempting to speculate about the extent to which these four properties might characterize successful social behavior in competitive situations. Axelrod's book offers some provocative generalizations, and we will look again at these ideas in Chapter 15.

Although Prisoner's Dilemma is the most famous 2×2 non-zero-sum game, there are other games which seem to capture other interesting aspects of social interaction. Some of them are presented in the Exercises. Prisoner's Dilemma and related games have been widely used in experimental work in social psychology. Chapter 13 gives one example of such use.

Exercises for Chapter 12

1. Suppose Rose and Colin play Game 12.1 twice in succession.
   a) Each player has eight strategies in this "supergame." List the eight Rose strategies, and devise a good notation for them.
   b) Write the 8×8 matrix for this game.
   c) Two Rose strategies together dominate all other Rose strategies (and similarly for Colin). Find these two strategies and say what would happen in the resulting 2×2 game.
   d) Compare your analysis in c) with the verbal argument in the text about repeated play of Prisoner's Dilemma.

2. a) Show that for any number $p \neq 1$, $\quad 1 + p + p^2 + \cdots + p^{m-1} = \dfrac{1 - p^m}{1 - p}$.

   b) Argue that if $0 \leq p < 1$, then $\quad 1 + p + p^2 + p^3 + \cdots = \dfrac{1}{1 - p}$.

3. Game 12.4 has two Nash equilibria besides Rose XII-Colin II. Find them. Are they cooperative or non-cooperative?

4. Game 12.5 is known as "Chicken," after a teenage game popular in the 1950's. Two cars drive straight toward each other at high velocity, and the first driver to swerve loses face. A is "swerve," B is "don't swerve."

		Colin	
		A	B
Rose	A	(0,0)	(−2,1)
	B	(1,−2)	(−8,−8)

Game 12.5: Chicken

   a) Find the pure strategy Nash equilibria in this game. Are they Pareto optimal?
   b) There is also a mixed strategy Nash equilibrium. Find it.
   c) How would you play Chicken? What would you do if you were going to play it 10 times? (Some results on what other people do are in [Rapoport and Chammah, 1969].)

5. Rose and Colin are going to meet at either the artfair or the ballgame, but they left home without deciding which. Rose prefers the artfair and Colin prefers the ballgame. Of course, they would strongly prefer to meet rather than not meet, and it's too late to get in touch. Construct and analyze a game which represents their dilemma.

6. a) Think of two situations other than those mentioned in the text, which could be modeled by Prisoner's Dilemma. Say specifically who the players are, what the strategies are, and why the payoffs meet the Prisoner's Dilemma conditions.
   b) Do the same for two situations which could be modeled by Chicken (Game 12.5).

13. Application to Social Psychology: Trust, Suspicion, and the F-Scale

As a central part of their monumental study *The Authoritarian Personality* in 1950, T. W. Adorno and his co-workers developed one of the prototypes of the personality inventories now so familiar to us all. The F-scale inventory was a series of statements to which subjects were to respond by writing a number from 1 (strongly disagree) to 7 (strongly agree). The statements were designed to test personality variables which, the authors argued, underlay susceptibility to authoritarian ideologies. Here are the personality variables included, together with corresponding sample items from the inventory:

Conventionalism: "One should avoid doing things in public which appear wrong to others, even though one knows that these things are really all right."

Authoritarian submission: "What this country needs is fewer laws and agencies, and more courageous, tireless, devoted leaders whom the people can put their faith in."

Authoritarian aggression: "Homosexuality is a particularly rotten form of delinquency and ought to be severely punished."

Anti-introspection: "There are some things too intimate or personal to talk about even with one's closest friends."

Superstition and stereotyping: "Although many people may scoff, it may yet be shown that astrology can explain a lot of things."

Power and toughness: "Too many people today are living in an unnatural, soft way; we should return to the fundamentals, to a more red-blooded, active way of life."

Destructiveness and cynicism: "When you come right down to it, it's human nature never to do anything without an eye to one's own profit."

Projection and exaggerated sexual concern: "The sexual orgies of the old Greeks and Romans are nursery school stuff compared to some of the goings-on in this country today, even in circles where people might least expect it."

The F-scale was controversial from its beginning, and has been the subject of considerable research in social psychology. For some of the issues, see [Christie and Jahoda, 1954]. In the research we will be concerned with, Morton Deutsch [1958, 1960] investigated the connection between Adorno's "authoritarian personality" and the commonsense concepts of trust, suspicion, and trustworthiness. Would it be true that people who scored high on the F-scale would be more suspicious of the behavior of others, and less trusting and trustworthy? The first problem was to make the concepts of trust, suspicion, and trustworthiness operational—to design an experimental situation in which we can identify specific types of behavior as showing trust or suspicion, or being trustworthy. Deutsch's idea was to use a Prisoner's Dilemma game to do this.

Deutsch first had his subjects (students in an introductory psychology class) fill out the F-scale questionnaire. Several weeks later he had them play the Prisoner's Dilemma game

		Second player	
		A	B
First player	A	(9, 9)	(−10, 10)
	B	(10, −10)	(−9, −9)

The first player chose A or B, and her choice was announced to the second player. Then the second player made his choice, and the payoffs were announced. Each subject played the game twice, once as first player and once as second player, with different (and unknown) partners each time.

Notice that if the first player chooses A, she guarantees a large positive payoff to the second player. The second player can then choose A and give the first player a large positive payoff in return, or choose B and get slightly more for himself while giving the first player a large negative payoff. If the first player chooses B, the outcome will presumably be -9 for both players. What the first player does, therefore, will depend on what she thinks the second player will do if she chooses A. If she thinks that the second player will choose A, she should choose A. If she thinks that the second player will choose B if she chooses A, she should "cut her losses" by choosing B. Deutsch used this analysis to obtain the following operational definitions:

- *Trust* is choosing A when you are the first player.
- *Suspicion* is choosing B when you are the first player.
- *Trustworthiness* is choosing A when you are the second player, and the first player has chosen A.
- *Untrustworthiness* is choosing B when you are the second player, and the first player has chosen A.

Do you agree these are reasonable? In the actual experiment, when subjects played as the second player, the first player was fictional—the experimenter always announced that she had chosen A.

Deutsch's first finding was that there is a strong correlation between being trusting and being trustworthy:

	Trustworthy	Untrustworthy	
Trusting	24	5	$\chi^2 = 24.7$
Suspicious	4	22	$p < .001$

Subjects who were trusting as the first player were in general trustworthy as the second player, and subjects who were suspicious as the first player were correspondingly untrustworthy as the second player. As Deutsch put it,

> It is apparent that the subjects were not behaving in accordance with the ethical injunction "Do unto others as you would have others do unto you," but were rather guided by the dictum of cognitive consistency, "Do unto others as you expect them to do unto you, and expect others to do unto you as you do unto them." [Deutsch, 1960]

To relate this behavior to the F-scale, Deutsch divided F-scale scores into categories of low (1.2–2.2), medium (2.3–3.3) and high (3.4–4.4). Notice that since the maximum possible score is 7, "high" only means relatively high with respect to a generally non-authoritarian population. Here are the results:

	F-scale score			
	Low	Medium	High	
Trusting and trustworthy	12	10	2	
Suspicious or untrustworthy	2	7	0	$\chi^2 = 23.6$
Suspicious and untrustworthy	0	13	9	$p < .005$

There is a strong correlation between being suspicious and untrustworthy, and scoring relatively high on the F-scale.

We need to be careful what we conclude from this result. What is certainly true is that we have discovered a behavioral correlate of the F-scale: relatively high scorers will, in general, play a sequential Prisoner's Dilemma game differently from low scorers. It also seems reasonable to interpret ways of playing this game as trusting or suspicious or trustworthy. However, there are a number of alternative explanations we might give for the correlation we have discovered. Deutsch notes two.

One possible explanation, consistent with the original psychoanalytic conclusions of Adorno and his colleagues, is that scoring high on the F-scale indicates a particular kind of personality structure in which "the conscience or superego is incompletely integrated with the self or ego," which results in the person being suspicious and untrustworthy. However, a different explanation might be that relatively high scores on the F-scale represent values that have been learned from unfortunate experience. A person has found others to be untrustworthy, has reacted by being suspicious, and this justified suspicion has contributed to forming the values measured by the F-scale.

The difference in possible interpretations of this experiment should warn us that experimental games cannot take the place of careful thought in social psychology. However, experimental games can give psychologists a way to make previously vague concepts precise and operational, and provide measurable results about connections between those concepts. Since the 1950's there has been a large amount of research involving experimental games, with Prisoner's Dilemma playing a central role. Anatol Rapoport estimated that 200 experiments involving Prisoner's Dilemma were reported in the literature between 1965 and 1973. By 1982, Andrew Colman's estimate was 1000. [Rapoport, 1974] and Chapter 7 of [Colman, 1982] are good places to start if you would like to explore some of this literature.

14. Strategic Moves

In our analysis of non-zero-sum games up to this point, we have required that the players choose their strategies simultaneously and not communicate with each other beforehand. Of course, games in real life may not be like this. In this chapter we will consider some of the things which might happen when one player can move first and make his move known to the other player, or when the players can talk to each other before they move. Commitments, threats, and promises become possible. The classic analysis of these kinds of "strategic moves" is [Schelling, 1960], which I strongly recommend.

To begin, let us consider the effect of moving first in a two-person game. Specifically, if you must move first, does that help you or hurt you? We will look at some examples.

		Colin	
		A	B
Rose	A	(3, −3)	(0, 0)
	B	(−1, 1)	(4, −4)

Game 14.1

Game 14.1 is zero-sum. Optimal strategies are Rose $(\frac{5}{8}A, \frac{3}{8}B)$ and Colin $(\frac{1}{2}A, \frac{1}{2}B)$ and the value of the game is $\frac{3}{2}$ to Rose. If Rose moves first, then Colin's choice can be contingent on Rose's move. If Rose chooses A, Colin will choose B and the payoff will be 0. If Rose chooses B, Colin will choose A and the payoff will be -1. Rose would prefer 0 to -1 and so will choose A, and the outcome will be AB $= (0, 0)$. The privilege of moving first has cost Rose an expected payoff of $\frac{3}{2}$. In fact, it is clear if you think about it that in a zero-sum game, where interests are completely opposed, it can only hurt you to move first. This is by no means true, though, in non-zero-sum games. A prototypical example is the game of Chicken.

		Colin	
		A	B
Rose	A	(3, 3)	(2, 4)
	B	(4, 2)	(1, 1)

Game 14.2: Chicken

In Game 14.2 whichever player moves first can choose B. It is then in the other player's best interest to choose A, and the player with the first move has obtained his most desired outcome.

In Chicken, both players would like to move first; in a zero-sum game both players would like the other player to move first. That doesn't exhaust the possibilities.

		Colin A	Colin B
Rose	A	(2, 3)	(4, 1)
	B	(1, 2)	(3, 4)

Game 14.3

In this game Rose A dominates Rose B, so we would expect the outcome to be AA if the players move simultaneously. AA is an equilibrium, but it is not Pareto optimal. If Colin moves first, you can check that the outcome would still be AA. However, suppose Rose moves first. Rose reasons that if she chooses A, Colin will choose A and the outcome will be AA; if she chooses B, Colin will choose B and the outcome will be BB. Since Rose prefers 3 to 2, she will choose B, giving the Pareto optimal outcome BB and helping *both* players. In this game, both players would like for Rose to move first.

If it is not possible for one player to move first in a game, the same effect can be obtained if the players can communicate and one player has a way to make a *commitment* to a move. For instance, if Rose can commit herself to strategy B in Chicken, then Colin can do no better than to choose A and give Rose her best possible outcome. Of course, the problem is how to make a commitment convincing to the other player, especially when the other player would also like to commit and when, as in Chicken, conflicting commitments are damaging. Schelling has a number of interesting suggestions. One is to make a commitment ("I am going to choose B.") and then cut off communication (hang up the phone, and take it off the hook). The other player then has the choice of giving in, or getting an even worse outcome (at BB in Chicken). We'll see another method of commitment a little later.

Communication can take forms other than attempts to make commitments to a first move.

		Colin A	Colin B
Rose	A	(4, 3)	(3, 4)
	B	(2, 1)	(1, 2)

Game 14.4: Threat-vulnerable equilibrium

With simultaneous moves Game 14.4 is solvable in the strict sense, having a unique equilibrium at AB, which is Pareto optimal. You can check that the result is still AB if either Rose or Colin moves first. Thus commitments would not affect the outcome of the game. However, suppose Colin is to move first and

Rose makes the *threat*: "If you choose B, then I will choose B." We call this a threat because it has the following properties:

i) Rose says that she will take a certain action contingent on a previous action by Colin.
ii) Rose's action will be harmful to Colin.
iii) Rose's action will also be harmful to Rose.

If Colin believes Rose's threat, then his choice is between BB if he chooses Colin B, or AA if he chooses Colin A. Colin would prefer AA, so this would be the outcome. Rose's threat, if believed, would give Rose her highest payoff.

Of course, the principal problem with a threat, in addition to any moral or ethical difficulties, is that it must be credible, and this is difficult because of condition iii). The threat is designed to deter Colin from his natural course of action (choosing B in our example), but if Colin is not deterred, then Rose has no incentive to carry out her threat. To do so would hurt Rose *without* having any chance to change Colin's choice. How can you convince another player that you would make a choice harmful to yourself after it is too late for that choice to do any good?

Again, Schelling and others have some ideas. One of the most common applies to repeated games. If the game is to be played several times, it might be worth Rose's while to take a low payoff in the first few games in order to establish her credibility for later games. Colin, knowing this, might find Rose's threat credible in those early games. But what about games toward the end? It is also true that a threat is a commitment to a (contingent) action, so that techniques which make a commitment convincing might also help with threats. However, a threat is more difficult because the commitment is to a self-harmful action. We will return to this problem shortly.

There are games where threats will not work, but gentler techniques will.

		Colin	
		A	B
Rose	A	(3, 3)	(−1, 5)
	B	(5, −1)	(0, 0)

Game 14.5: Prisoner's Dilemma

Recall that in the Prisoner's Dilemma the natural outcome is at BB with simultaneous moves or with either player moving first. Furthermore, neither player has a threat in this game. If Colin goes first, for example, no matter which choice he makes, Rose is motivated to choose Rose B, the strategy most harmful to Colin. Rose can't threaten to do anything worse than she would do naturally! What is needed in Prisoner's Dilemma is not a threat, but a *promise*: "If you choose A, then I will choose A." This is a promise because it has the following properties:

i) Rose says that she will take a certain action contingent on a previous action by Colin.
ii) Rose's action will be beneficial to Colin.
iii) Rose's action will be harmful to Rose.

If Colin believes Rose's promise, his choice is between BB if he chooses B, and AA if he chooses A. He will choose AA and both players will benefit.

Of course a promise is subject to the same difficulty as a threat: it must be made credible, and this is hard because of property iii). Again, Rose is making a contingent commitment which she will have no incentive to carry out if Colin believes it. It is not even enough that both players would benefit if Colin believes the promise and Rose carries it out. Colin, wanting to believe, still knows that if he does, Rose will benefit by betraying him. Again, repeated play can help this situation, at least in the early games, since both players know that Rose could benefit by establishing credibility.

As a last example, consider this game with Colin moving first:

		Colin	
		A	B
Rose	A	(3,3)	(1,5)
	B	(4,0)	(0,2)

Game 14.6: Equilibrium vulnerable to a threat-and-promise

The natural outcome is at AB, giving Rose her next to lowest payoff. If Rose threatens "If you choose B, I will choose B," it will not affect Colin's action since Colin, having a choice between BA and BB, would prefer BB. Similarly, if Rose promises "If you choose A, I will choose A," it will not affect Colin's action. What is need here is a combination of *both* the threat and the promise. If Rose makes both credible, Colin will have the choice between AA if he chooses A, and BB if he chooses B. He will choose A, and Rose will benefit. Sometimes one needs both the carrot and the stick. There are also many games (see Exercise 2) in which no combination of threats and/or promises can change the outcome.

For all of the strategic moves—commitments, threats, promises—the major problem is making them credible. Schelling pointed out that many methods of establishing credibility are equivalent to *lowering one or more of one's own payoffs*. For example, consider Rose's threat in Game 14.4, "If you choose B, I will choose B." How could Rose convince Colin that she would really choose a payoff of 1 at BB instead of a payoff of 3 at AB? She might pledge her honor on carrying out the threat. She might call in witnesses and explain that if they saw her not carry out the threat, her word would never again be taken seriously. She might sign a legal contract to forfeit $1000 to a third party if she failed to carry out her threat. All of these maneuvers have the effect of lowering Rose's utility for AB once Colin has chosen B. If Rose can convince Colin that she has lowered this utility to below her utility level for BB, Rose's threat is credible.

Here is how the other strategic moves we have considered could be implemented by lowering payoffs:

- in Game 14.2 Rose could commit to Rose B by lowering her payoff at AB from 2 to 0.
- in Game 14.5 Rose could make her promise credible by lowering her payoff at BA from 5 to 2.
- in Game 14.6 Rose would have to lower her payoff at AB from 1 to -1 (the threat), and also lower her payoff at BA from 4 to 2 (the promise).

What these examples show is that, paradoxically, it can be an advantage in a game to be able to *lower* some of your own payoffs. (Of course, it would also be nice to be able to raise your own payoffs, or to affect the other player's payoffs, but this is usually more difficult to do!)

When a game is played several times, we have noted that players who can communicate can use the repeated play to try to establish the credibility of their commitments, threats or promises. Repeated play also gives players who *cannot* communicate a chance to make implicit strategic moves. For example, TIT FOR TAT in iterated Prisoner's Dilemma could be interpreted as an operational way of conveying the promise "If you cooperate, so will I," along with the (strategically unnecessary but psychologically useful) threat "If you don't cooperate, neither will I." In fact, strategic moves can be conveyed quite effectively in iterated play even if the players do not know each others' payoffs.

Table 14.1 shows the payoffs in a game adapted from [Harnett, 1967] which students in my game theory classes have played and found enlightening. One person is a wholesaler who sets the price at which a retailer can buy a certain item. The other person is the retailer who, once the price is announced, places an order for how many items she wishes to buy at that price. This determines the profit of both players, as shown in the table. Notice that if the price is low the retailer can benefit by placing a large order, but if the price is high the retailer should not place a large order, since she will not be able to sell many items at a high price. Since the wholesaler is selling to just one retailer and the retailer is buying from just one wholesaler, this situation is known as a "bilateral monopoly."

The Wholesaler's Game

		Wholesaler's price			
		$9	$10	$11	$12
Retailer's order	0	0	0	0	0
	8	22	27	34	42
	16	32	42	56	70
	20	30	42	61	78
	28	26	43	69	93

(payoffs to wholesaler)

The Retailer's Game

		Wholesaler's price			
		$9	$10	$11	$12
Retailer's order	0	0	0	0	0
	8	28	23	16	8
	16	48	38	24	10
	20	54	42	23	6
	28	62	45	19	−5

(payoffs to retailer)

▭ Bowley points ┌╴ Resistance points ▯ Equal profits points

Table 14.1: Price leadership in a bilateral monopoly.

In the class experiment, the students were divided into 13 pairs. The 13 wholesalers were in one room, the 13 retailers in another, so they could not communicate. The wholesalers had only their payoff matrix and the retailers had only theirs. Each pair played the game 15 times. The wholesalers would choose prices, I would announce those prices to the retailers, the retailers would choose their orders, and I would carry those back to the wholesalers. Remember that each wholesaler is dealing with just one retailer and visa versa, but they all did have a chance to see what was happening with other pairs. What would you guess would happen?

If the players knew each others' payoffs and the game were played just once, with the wholesaler moving first, the rational outcome would be a price of $12, an order of 16, and payoffs of (10, 70). This outcome is called in economic theory the *Bowley point* of the game. In the experiment, many players did play either the Bowley point or the nearby (24, 56), a kind of "second best Bowley point." In the analysis, I'll call them both Bowley points.

Another common outcome was that the wholesaler would set a price of $11 or $12 and the retailer would place an order of 0, or sometimes 8, in an attempt to make the wholesaler lower the price in the next round. I'll call these outcomes "resistance points." Finally, if the players could see each others' payoffs and were trying to be fair, they might be attracted to the "equal profits point" (42, 42) or its Pareto optimal modification (45, 43). I have marked these different types of points on the table.

In the play, most wholesalers started by setting prices of $11 or $12. Most retailers went along for a couple rounds, but then began to resist, placing orders of 0 to communicate the threat, "If you keep your price high, I won't buy." The public nature of the proceedings had interesting effects, with retailers encouraging each other to resist. The wholesalers' responses to resistance varied. Some gave in and lowered their price, either immediately or after several rounds. Others kept the price high or even raised it from $11 to $12 to communicate "I'm committed to a high price and I can't be intimidated." At this point the language in both rooms began to get quite spicy. When wholesalers didn't give in, sometimes the retailers gave up resisting and returned to the Bowley point. Sometimes mutual resistance lasted all the way to the end of the 15 games. By the end of the experiment, all pairs had settled into stable patterns. Five of them were at a Bowley point, four at an equal profits point, and four at a resistance point. Which point they settled at had a lot to do with the total payoffs (over all 15 games) they received:

Type of final outcome	Average wholesaler payoff	Average retailer payoff
Bowley points (5)	656	242
Equal profits points (4)	513	484
Resistance points (4)	263	160

Perhaps the most interesting result of the experiment was the ease and subtlety with which threats and counter-threats could be conveyed and interpreted with no overt communication, and the diversity of responses to those threats. In one third of the cases the retailer's original threat worked, in one third of the cases it met strong resistance and was eventually withdrawn, and one third of the pairs locked in on the mutually harmful course of mutual resistance.

After the original fifteen rounds, I gave both players complete information about each others' payoffs, and had them play five more rounds to see if complete information would change the outcome. In 10 of the 13 pairs it did not; the psychological set was too strong. One pair changed from a Bowley point to an equal profits point, one pair from a resistance point to a Bowley point, and one pair which had been at a Bowley point worked out a clever alternation between Bowley and equal profits points. For more results, you might enjoy consulting [Fouraker and Siegel, 1960] and [Siegel and Harnett, 1964].

Since the work of Schelling on commitments, threats and promises, a complete analysis of which 2×2 games are vulnerable to these strategic moves has been carried out in [Rapoport and Guyer, 1966, 1976]. In Exercise 4, you are invited to see how they classified 2×2 games into 78 ordinal types.

Exercises for Chapter 14

1. The following games all have outcomes which do not give Rose her maximal payoff. For each one, say whether Rose could benefit by any of the following strategic moves: i) seizing (or committing to) the first move, ii) forcing Colin to move first, iii) making a threat, iv) making a promise, or v) making both a threat and a promise. For each game, it is possible that more than one of these could work, or that none of them will. For each commitment, threat or promise, show how Rose could implement it by lowering some of her own payoffs.

a)

		Colin	
		A	B
Rose	A	(3,4)	(4,3)
	B	(2,2)	(1,1)

b)

		Colin	
		A	B
Rose	A	(3,4)	(4,2)
	B	(2,3)	(1,1)

c)

		Colin	
		A	B
Rose	A	(2,4)	(3,3)
	B	(1,2)	(4,1)

d)

		Colin	
		A	B
Rose	A	(2,2)	(4,1)
	B	(1,3)	(3,4)

e)

		Colin	
		A	B
Rose	A	(3,2)	(1,1)
	B	(2,4)	(4,3)

2. Make up a 2×2 game, different from any in Exercise 1, which is solvable in the strict sense and whose solution cannot be affected by any commitment, threat and/or promise.

3. A kidnapper takes a hostage and demands that the hostage pay ransom. They then play the following game in extensive form. The hostage may pay the ransom or not. Following that, the kidnapper may kill the hostage or release him. If the hostage is released, he may report the kidnapping to the police or not report it. The kidnapper has utility $+5$ for getting paid, -2 for having the kidnapping reported, and -1 for killing the hostage. These utilities are additive (for instance, getting paid but also reported is worth $5 - 2 = 3$). Utilities for the hostage are -10 for getting killed, -2 for paying, $+1$ for reporting, and these are also additive.
   a) Draw the game tree for this game, labeling choosers, choices and payoffs.
   b) Find the natural outcome of the game. [It should be the second worst outcome for both players.]
   c) Suppose the hostage could make a convincing threat or promise. What should it be? What is the new outcome? Who benefits?
   d) Suppose that after the hostage makes the convincing threat or promise in c), the kidnapper can make a threat or promise. What should it be? What is the new outcome? Who benefits?
   e) If the hostage is not able to make a convincing threat or promise, but the kidnapper can, what should it be and what is the result?
   f) Suggest some real-life ways that any or all of the above threats and promises might be made credible.

4. [Rapoport and Guyer, 1966] showed that there are exactly 78 distinct ordinal types of 2×2 games, where two games are not considered distinct if they differ only by how the strategies are labelled, or if one can be obtained from the other by interchanging the players. Verify this as follows:
   a) Show that there are 576 ways in which Rose and Colin ordinal utilities 1, 2, 3, 4 can fill in a 2×2 matrix.
   b) If we always label "A" the Rose strategy which contains Rose's payoff 4, and do the same for Colin, the number of games is reduced to 144.
   c) Of these 144 games, 12 are symmetric (i.e. would not be changed by interchanging the players).
   d) Hence the number of distinct games is $12 + (132/2) = 78$.

15. Application to Biology: Evolutionarily Stable Strategies

The idea of an evolutionarily stable strategy (ESS), first introduced by John Maynard Smith and G. R. Price [1973], is a powerful explanatory idea in evolutionary biology. It is especially applicable to the study of behavior, and has found an important place in modern sociobiology. The basic idea is this. Because individual members of a biological species have similar needs, and resources are limited, conflict situations will often arise. In these conflict situations, there are many different behavior patterns (strategies) which individuals might follow. Which ones will they choose? Since the question involves behavior in conflict situations, a game-theoretic formulation should be helpful. On the other hand, because birds or fish or insects do not calculate rationally, game-theoretic solution ideas based on rational behavior will probably not be relevant. What takes the place of rationality is evolutionary pressure—natural selection. The solution concept it leads to is that of an ESS.

We will start with a simple conflict model proposed by Maynard Smith and Price, and discussed in [Dawkins, 1976]. Members of a species will engage in repeated random conflicts over some resource. Each conflict is between two members, only one of whom can win the resource. Winning the resource is worth 50 "fitness points." Fitness points are to be interpreted as increased probability of passing along genes to the next generation. In a first model, we suppose that an individual has only two possible strategies, labeled by the suggestive names of "hawk" and "dove." A hawk fights for the resource; a dove merely engages in symbolic conflict, posturing and threatening but not actually fighting. If both players adopt the hawk strategy, they will fight until one is injured. The winner will get the resource worth 50 points, while the loser will get -100 points for being injured. If a hawk meets a dove, the hawk will always win the resource and there will be no injury. If two doves meet, they will spend a long time posturing. One of them will eventually win the resource, but they will both get -10 points for wasted time. The payoff matrix is

		Player Two	
		hawk	dove
Player One	hawk	$(-25, -25)$	$(50, 0)$
	dove	$(0, 50)$	$(15, 15)$

Game 15.1

The hawk-hawk payoffs, for example, are expected values: $\frac{1}{2}(50) + \frac{1}{2}(-100) = -25$. Of course, the specific numbers we have chosen are not really important: it is the qualitative ideas which matter. Notice that this game is non-zero-sum, that the same strategies are available to both players, and that the payoffs to the players are symmetric. Although ESS analysis is also useful in non-symmetric situations (Exercise 6), we will deal here only with symmetric games. For symmetric games, it is sufficient to give only the payoffs to Player One, for the payoffs to Player Two can then be filled in symmetrically.[†]

We are interested in how a species of members who play these strategies will evolve. The basic assumption is that behavior in this kind of conflict situation is genetically influenced. The simplest form of that assumption is that hawk or dove behavior is genetically determined, so that each individual either always plays hawk or always plays dove, and tends to pass this behavior on to his progeny. We can make this strong assumption for simplicity, but we'll see that it is not really necessary.

Suppose that the population starts off being almost entirely doves. Any individual in this population plays almost all of his games against doves. A dove could therefore expect an average of 15 fitness points per game, whereas a rare hawk could expect an average of 50 points. Since the hawks are genetically advantaged, more of the next generation would be hawks, and over the course of time the hawk population would rise. A population of doves is not *evolutionarily stable*: a small hawk minority (which might arise, for example, by mutation) would increase and invade such a population. Similarly, a population of hawks would not be evolutionarily stable, for in such a population doves would be advantaged, winning an average of 0 points per encounter compared to an average of -25 points for hawks.

What about a mixed population, for instance $\frac{1}{4}$ hawks and $\frac{3}{4}$ doves? We can figure out which strategy would be best in this environment by thinking of a "focal player" playing against an opponent who uses a mixed strategy of $\frac{1}{4}$ hawk, $\frac{3}{4}$ dove, and doing an expected value calculation:

		Other Player: hawk	Other Player: dove	Expected payoff to Focal Player
Focal Player	hawk	-25	50	$(\frac{1}{4})(-25) + (\frac{3}{4})(50) = 31\frac{1}{4}$
	dove	0	15	$(\frac{1}{4})(0) + (\frac{3}{4})(15) = 11\frac{1}{4}$
Other Player strategy:		$\frac{1}{4}$	$\frac{3}{4}$	

In this situation it pays to be a hawk, and hawks would increase.

We have seen that if there are few hawks, hawks will increase, whereas if there are few doves, doves will increase. There should be a proportion of hawks and doves for which these two tendencies balance out. To find it, we calculate the

[†]Notice that the situation here is different from that of a zero-sum game, where the payoffs to Player Two are the negatives of the payoffs to Player One.

mixed strategy which, when played by the other player, leaves the focal player indifferent between playing hawk and playing dove. In that situation neither hawks nor doves would be advantaged and the proportion should stay fixed.

		Other Player		
		hawk	dove	Expected payoff to Focal Player
Focal Player	hawk	−25	50	$(-25)x + 50(1-x) = 50 - 75x$
	dove	0	15	$(0)x + 15(1-x) = 15 - 15x$
Other Player strategy:		x	$1-x$	

Setting $50 - 75x = 15 - 15x$, we solve for $x = \frac{7}{12}$. Thus a population of $\frac{7}{12}$ hawks and $\frac{5}{12}$ doves would be evolutionarily stable. We say that the mixed strategy of $\frac{7}{12}$ hawk, $\frac{5}{12}$ dove is an *evolutionarily stable strategy (ESS)* for this game.

At this point, we can notice that we don't need the assumption that each individual is either a pure hawk or a pure dove. The same kind of stability would be obtained if all individuals played a mixed strategy of $\frac{7}{12}$ hawk, $\frac{5}{12}$ dove and this propensity were genetically passed on to future generations. Or different individuals could play different mixed strategies, averaging to $\frac{7}{12}$ hawk, $\frac{5}{12}$ dove across the population. Mathematically, the results are the same.

The general condition for a strategy S to be an ESS is this:

- Let T be any other strategy, pure or mixed. Suppose that everyone in the population is playing S, except for a few players who are playing T. Then the expected payoff for playing S must be at least as great as the expected payoff for playing T.

This condition says exactly that if a population has adopted S, no mutant strategy T could invade and prosper against S. We would expect evolution to produce evolutionarily stable strategies. The argument is that a strategy which is not an ESS would have been invaded by another strategy. In the long run, only ESS's should survive. We therefore have a powerful theoretical tool for explaining behavior on evolutionary grounds: surviving behavior patterns, if they are genetically influenced, should be ESS's in an appropriate game.

We have seen from the hawk-dove example that an ESS may be a mixed strategy. It can also be a pure strategy, and there can be more than one ESS in a particular game. Consider the following examples.

		Player Two	
		A	B
Player One	A	1	2
	B	3	4

Game 15.2

		Player Two	
		A	B
Player One	A	3	1
	B	2	4

Game 15.3

		Player Two	
		A	B
Player One	A	1	4
	B	2	3

Game 15.4

In Game 15.2, strategy B is an ESS, and is the only ESS. One easy way to see this is to note that B strictly dominates A, so that in any population, B is advantaged. In Game 15.3, both A and B are ESS's. In a population of almost all B's, B would be best (by 4 to 1); in a population of almost all A's, A would be best (by 3 to 2). Whichever strategy became established first would persist. Finally, in Game 15.4 neither pure strategy is an ESS. The unique ESS is a mixed strategy. In fact, Game 15.4 is ordinally equivalent to our original hawk-dove Game 15.1.

[Maynard Smith and Price, 1973] noted the following conditions for pure strategy ESS's in a general 2×2 symmetric game

		Player Two: A	Player Two: B
Player One	A	a	b
	B	c	d

(payoffs to Player One)

A is an ESS if $a > c$, or $a = c$ and $b \geqslant d$.
B is an ESS if $d > b$, or $d = b$ and $c \geqslant a$.

Game 15.5

If neither condition holds, there is a unique mixed strategy ESS which can be found as we did for Game 15.1 (Exercise 1).

In a game with more than two strategies, a pure strategy S is an ESS if the diagonal entry for S is the largest entry in its column, since then S is the best strategy to play when almost everyone is playing S. (If the diagonal entry for S ties for being the largest entry in its column, further checks are necessary.)

The definition of an ESS is that it must be able to resist invasion by any other strategy T. This gives rise to a major difficulty in applying the idea of an evolutionarily stable strategy in practical contexts. We must be sure that we have identified all feasible strategies in a game, if we are to be sure that a given strategy is an ESS. To illustrate this problem, let us consider two variations of the hawk-dove game, also due to Maynard Smith and Price. Recall that if only the hawk and dove strategies (or mixtures of them) are available to the players, then $\frac{7}{12}$ hawk, $\frac{5}{12}$ dove is the unique ESS. But let's expand our horizons. Since it is advantageous to play hawk against a dove, and advantageous to play dove against a hawk, why not design a conditional strategy which would do just that? In fact, the following strategy is even slightly better.

- **Bully:** In any contest, show initial fight. Continue to fight if your opponent does not fight back. If your opponent fights back, run away.

If we introduce this new possibility, the game looks like this:

Player Two

		hawk	dove	bully
Player One	hawk	−25	50	50
	dove	0	15	0
	bully	0	50	25

(payoffs to Player One)

Game 15.6

I think the payoffs are clear, except possibly for what happens when two bullies meet. The assumption there is that both of them will want to run away, but one will run away faster and the other will be left holding the prize. You can check that in this game no pure strategy is an ESS. To find an ESS, notice that the bully strategy dominates the dove strategy, so that doves would die out. Hence the game reduces to only hawks and bullies, and the mixed strategy $\frac{1}{2}$ hawk, $\frac{1}{2}$ bully is an ESS (and the only ESS). The prognosis seems to be discouraging: it looks like we can expect a lot of conflict and cowardice.

However, before we can reach this conclusion, we have to sure we haven't left out other feasible conditional strategies. In fact, the relevant question is one most children have to deal with: what is the best way to deal with bullies? Perhaps as a child you found the answer, in something like the following strategy.

- **Retaliator:** In any contest, behave like a dove at first. However, if you are persistently attacked, fight back with all your strength.

Notice that a retaliator will always win the resource from a bully without a fight. Adding this new strategy, we get

Player Two

		hawk	dove	bully	retaliator
Player One	hawk	−25	50	50	−25
	dove	0	15	0	15
	bully	0	50	25	0
	retaliator	−25	15	50	15

(payoffs to Player One)

Game 15.7

In this game the pure strategy retaliator is an ESS. So is any mixture of retaliators and doves which contains less than 30% doves (with more than 30% doves, bullies would invade—Exercise 4). This result is a little paradoxical: retaliators, who behave like hawks against hawks and like doves against doves, would seem to have the worst of both worlds, yet in the course of evolution they come to dominate the population. The result is also biologically quite interesting. Notice that in a population of retaliators, we would see no fighting, but only posturing and symbolic conflict. Ethologists, for instance Konrad Lorenz in *On Aggression*, have noted this kind of behavior in many species. The game-theoretic analysis suggests that this kind of peaceful equilibrium can only be maintained by a willingness to fight. A population of doves is not evolutionarily stable; a population of retaliators is.

The alert reader will notice that we have been jumping too quickly to conclusions. To be sure such conclusions are justified, we should consider other conditional strategies, in fact all other conditional strategies, to see if any of them could upset the equilibrium of retaliators. In fact, however, the retaliator strategy is remarkably robust. To do better against a retaliator than another retaliator would do, you would have to win the resource from him without wasting time or risking injury. This is difficult because a retaliator will not run away, and if you fight him, he will fight back.

Finally, let's consider one different kind of modification of the conflict model which points to another important area of modern ecological thought. Return to the original hawk-dove game with its ESS of $\frac{7}{12}$ hawk, $\frac{5}{12}$ dove. We know that in this equilibrium population the average payoff to hawks and doves is the same. If we calculate the size of this payoff we get

$$\tfrac{7}{12}(-25) + \tfrac{5}{12}(50) = 6\tfrac{1}{4} = \tfrac{7}{12}(0) + \tfrac{5}{12}(15).$$

All members of this population win, on average, $6\frac{1}{4}$ points per contest. Notice that this outcome is not Pareto optimal. For example, all members would be better off if everyone always played dove, for then everyone would win an average of 15 points per contest. Evolution can produce Pareto inferior outcomes.

Let us think for a moment about how well our players might do if they could somehow cooperate. The payoff polygon for the hawk-dove game is shown in Figure 15.1. The Pareto optimal symmetric outcome would be $\frac{1}{2}$DH, $\frac{1}{2}$HD with an average payoff of 25 to each player. Is there any way this could be achieved? What is needed is some kind of coordinating signal which will resolve all conflicts without a fight by telling one player to be a hawk, the other to be a dove. Here are some possible signals:

- fight only if you are bigger than your opponent
- fight only if your tail is longer than your opponent's
- fight only if your opponent is a lighter color than you
- fight only if you are on your "home ground".

The first three require that individuals make judgments about variations in size, tail length, or color. Notice that while the first signal might be related to who is likely to win a fight if there is one, the second and third are probably irrelevant. Their tactical irrelevance does not at all impair their ability to serve as signals, and there is substantial evidence that these kinds of signals do indeed exist in nature. The fourth kind of signal is of a different kind, and is connected to the development of *territoriality* in a species.

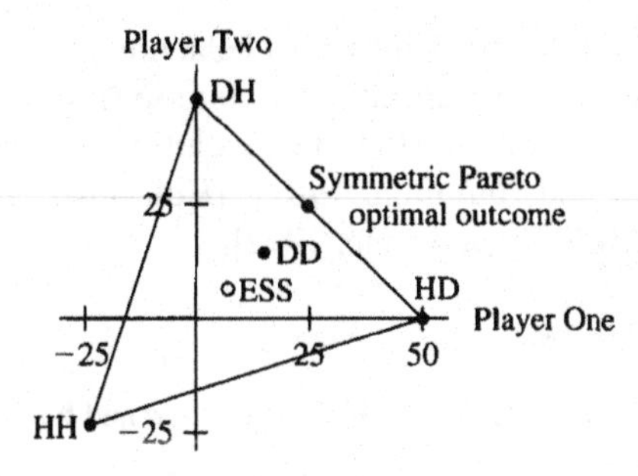

Figure 15.1: Payoff polygon for Game 15.1

If strategies involving coordinating signals are to develop and persist, it is not enough, of course, that they yield all members higher payoffs. The dove strategy is uniformly advantageous, but cannot persist because it is not evolutionarily stable. We must show that a coordinating strategy is an ESS when played against other strategies, including conditional ones. For our last game, we will try this for a territorial strategy:

- **Bourgeois:** be a hawk on your own territory, a dove on someone else's territory.

If we assume that a bourgeois will fight half of his contests on his own territory, and that two bourgeois are unlikely to meet on a third party's territory, the payoff matrix looks like

		Player Two				
		hawk	dove	bully	retaliator	bourgeois
Player One	hawk	−25	50	50	−25	12.5
	dove	0	15	0	15	7.5
	bully	0	50	25	0	25
	retaliator	−25	15	50	15	−5
	bourgeois	−12.5	32.5	25	−5	25

(payoffs to Player One)

Game 15.8

Since a bourgeois spends half his time as a hawk, half as a dove, the bourgeois payoffs are the average of the corresponding hawk and dove payoffs. However, when two bourgeois meet, territoriality decides the contest without a fight.

In Game 15.8 there are two different ESS's: retaliator (with some doves coexisting), and bourgeois (with some bullies coexisting). We might expect to see either kind of behavior.

We have been using game theory and the idea of an evolutionarily stable strategy to study the evolution of aggression. The ESS has been used in many other ways in biology, for instance to study sexual behavior, altruism, and cooperation. A simple sexual mating game appears in Exercise 6. For an account of work using the Prisoner's Dilemma to study the evolution of cooperation, I strongly recommend [Axelrod and Hamilton, 1981], [Hofstadter, 1983] and [Axelrod, 1984], and the account in [Poundstone, 1992].

Exercises for Chapter 15

1. In the general 2×2 symmetric Game 15.5 with $c > a$ and $b > d$, show that the mixed strategy ESS is

$$\frac{b-d}{(c-a)+(b-d)}\text{A}, \ \frac{c-a}{(c-a)+(b-d)}\text{B}.$$

2. In a 2×2 symmetric game, there are 24 ways to arrange the numbers 1, 2, 3, and 4 as payoffs in the game. If we agree to make the diagonal entry at BB larger than the diagonal entry at AA (which is just a convention on the labeling of strategies), we reduce the number of possibilities to 12. These give the 12 ordinal types of 2×2 symmetric games. Write out the 12 types (Games 15.2, 15.3, and 15.4 are three of them). Classify the 12 types by saying whether

   - only B is an ESS
   - only A is an ESS (this ESS will be non-Pareto-optimal)
   - both A and B are ESS's
   - the game has only a mixed ESS.

   Prisoner's Dilemma and Chicken are symmetric games. Identify them in your classification.

3. Verify that $\frac{1}{2}$ hawk, $\frac{1}{2}$ bully is an ESS in Game 15.6.

4. I claimed that in Game 15.7, bullies would invade a population of retaliators and doves if the fraction of doves was above $\frac{3}{10}$. Verify this by considering a population with a fraction x of doves, $(1-x)$ of retaliators, and calculating the expected payoffs of the four strategies in this population. Show that for $x > \frac{3}{10}$, the expected payoff for bullies is larger than for doves or retaliators. What about hawks?

5. Redo the analyses of Games 15.1, 15.6, and 15.7 if the resource in question is worth 150 points. The interesting change here is that the benefit of the resource now outweighs the cost of injury. Before you start, you might think about what effect this should have.

6. [Dawkins, 1976] considers a simple asymmetric mating game in which a female (a bird, say) tries to get a male to stay around and help raise a family of babies, instead of going off and propagating his genes elsewhere. One possible technique for doing this is to insist on a long and arduous courtship before mating. Suppose a female can be either *coy* (insist on courtship) or *fast* (be willing to mate with anyone), and a male can be either *faithful* (go through a courtship and then help raise the babies) or *philandering* (be unwilling to go through a courtship, and desert any female after mating). Suppose the payoff to each parent of babies is $+15$, and the total cost of raising babies is -20, which can be split equally between both parents, or fall entirely on the female if the male deserts. Suppose the cost of a long courtship is -3 to each player.

a) Verify that the resulting game is

		Male	
		Faithful	Philandering
Female	Coy	$(2, 2)$	$(0, 0)$
	Fast	$(5, 5)$	$(-5, 15)$

b) Draw the movement diagram of this game to show there is no pure strategy equilibrium.

c) A mixed strategy ESS for males would be one which equalizes the expected payoffs to coy and fast females. Find it.

d) Similarly, find an ESS for the females.

e) If males and females follow these ESS's, what will the expected payoffs be? Is this result Pareto optimal?

16. The Nash Arbitration Scheme and Cooperative Solutions

In our analysis of non-zero-sum games up to this point, the players have played non-cooperatively. Each has tried to do the best possible for himself by choosing strategies or by making strategic commitments, threats or promises. In this chapter we will consider a different approach. Imagine the players sitting down together to decide what is a reasonable or fair outcome to the game, and then agreeing to implement that outcome. Alternatively, imagine that the players call in an impartial outside arbitrator to determine a reasonable and fair outcome, and agree to abide by her decision. What principles should guide the players or the outside arbitrator in this context? Can we determine a reasonable and fair outcome to a game?

As an example, consider

		Colin A	Colin B
Rose	A	(2,6)	(10,5)
	B	(4,8)	(0,0)

Game 16.1

How would you advise the players to settle this game fairly? The most obvious advice might be for Rose and Colin to choose the outcome with the largest total payoff and then split that payoff equally. In this game they would play Rose A–Colin B for a total of 15, and split the payoff to get 7.5 each. Let us call this the *egalitarian proposal*. For all its democratic appeal, the egalitarian proposal is flawed in two different ways which illustrate important subtleties in our problem.

The first flaw is that the payoffs in the game are supposed to be utilities. Since we know from Chapter 9 that different players' utilities cannot meaningfully be added, we cannot calculate the "total utility" of each outcome to choose the largest total. Since different players' utilities cannot be compared or transferred, we cannot split utility equally. The egalitarian proposal just doesn't make sense.

The second flaw is that the egalitarian proposal, even if we could find a way to make sense of it, neglects the asymmetries of strategic position in the game. In this game Colin has a strong strategic position, since Colin A dominates Colin B and the natural outcome of the game is Rose B–Colin A, giving Colin his best outcome. Colin could reasonably argue that it is unfair to ignore the reality of his superior strategic position in determining a fair outcome to the game.

Our challenge is thus to find a method of arbitrating games which does not involve illegitimate manipulation of utilities, does take into account strategic inequalities, and has a claim to fairness. The first good idea along these lines goes back to von Neumann and Morgenstern [1944]. They argued that any reasonable arbitrated solution to a non-zero-sum game should be

i) Pareto optimal. There should not be another outcome which is better for both players, or better for one and equally good for the other.

ii) At or above the security level for both players. Neither player should be forced to accept less than he could guarantee himself by non-cooperative play.

The set of (pure and mixed) outcomes which satisfy these two conditions is called the *negotiation set* of the game. In Game 16.1 you can check that the security levels are $\frac{10}{3}$ for Rose, 6 for Colin. Looking at the payoff polygon in Figure 16.1, we see that the Pareto optimal outcomes are those on the line segment joining $(4,8)$ to $(10,5)$. Thus the negotiation set consists of outcomes on the line segment from $(4,8)$ to $(8,6)$. These correspond to Rose and Colin playing mixtures of BA and AB, with BA being played at least $\frac{1}{3}$ of the time. Notice that this prescription does not involve adding or comparing utilities, and it takes into account, via security levels, at least something of Colin's strategic advantage in the game. Of course, it only specifies a range of reasonable outcomes. Could we choose, within that range, a single outcome as fairest?

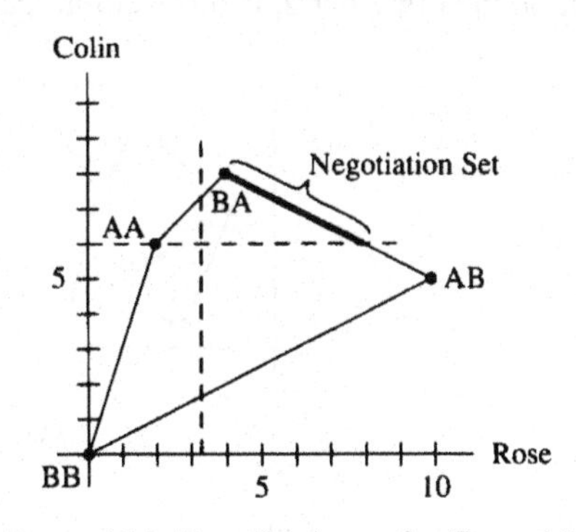

Figure 16.1: Payoff polygon for Game 16.1

A beautiful idea of how to do this was proposed by John Nash [1950b]. It is known as the *Nash arbitration scheme*. Nash considered a more general problem. Suppose two negotiating parties, Rose and Colin, have available to them a collection of outcomes which, when represented in a coordinate plane by their utilities to Rose and Colin, give a convex polygon. The parties try to agree on some outcome in this set. If they fail to agree, they will get some "default" outcome in the polygon, called the *status quo point*. How could they choose a fair point to agree on?

A general method to solve this problem must take any convex polygon in the plane, with a status quo point SQ in that polygon, and produce a point in the polygon as as solution. A method which does this will be called an *arbitration scheme*. Nash began by writing down four axioms (compare Chapter 10) which he believed a reasonable arbitration scheme should satisfy:

AXIOM 1: RATIONALITY. The solution point should be in the negotiation set.

AXIOM 2: LINEAR INVARIANCE. If either Rose's or Colin's utilities are transformed by a positive linear function, the solution point should be transformed by the same function.

AXIOM 3: SYMMETRY. If the polygon happens to be symmetric about the line of slope +1 through SQ, then the solution point should be on this line.

AXIOM 4: INDEPENDENCE OF IRRELEVANT ALTERNATIVES. Suppose N is the solution point for a polygon $\mathcal{P}$ with status quo point SQ. Suppose $\mathcal{Q}$ is another polygon which contains both SQ and N, and is totally contained in $\mathcal{P}$. Then N should also be the solution point for $\mathcal{Q}$ with status quo point SQ.

See Figure 16.2 for pictures of the axioms. The first three axioms seem quite straightforward. In particular, Axiom 3 embodies the idea of fairness as symmetry, or non-discrimination, in the special case of completely symmetric circumstances. Axiom 4, on the other hand, is more complicated. Nash argued for

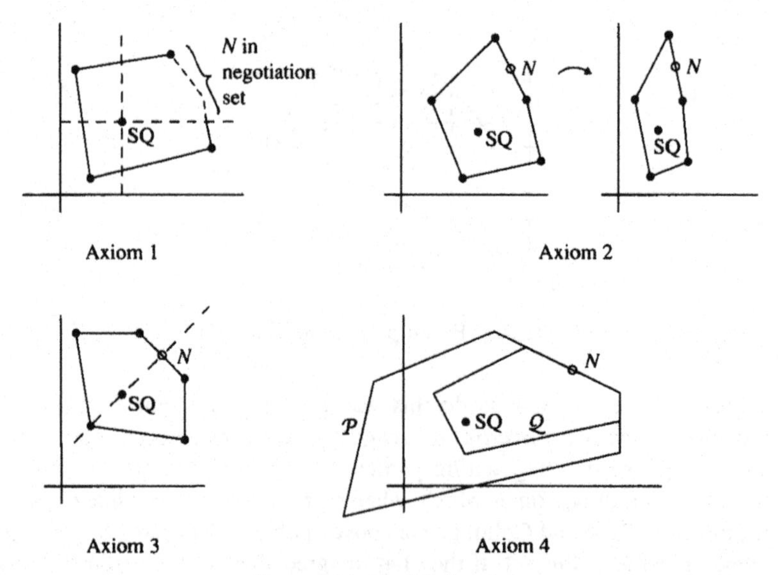

Figure 16.2: The Nash axioms

Axiom 4 as follows. Suppose in the situation where $\mathcal{P}$ is the set of available alternatives and SQ is the status quo, Rose and Colin agree that N is the fairest possible outcome. Now suppose that some of the outcomes in $\mathcal{P}$ are discovered to be not available after all—the set of available outcomes shrinks to $\mathcal{Q}$. Then N, which was judged fairer than all other outcomes in $\mathcal{P}$, should still be fairer than all other outcomes in $\mathcal{Q}$.

I find Nash's argument compelling, but we should be clear about the kind of things Axiom 4 rules out. In Figure 16.3, suppose that N is the solution to the arbitration problem (EAC, SQ). Then by Axiom 4 it must also be the solution to the problem (EAB, SQ). But if we thought that N was fair because it seemed to be a reasonable compromise between Rose's preferred point C and Colin's preferred point A, wouldn't a point like M be fairer if it turns out that C is not available and the best that Rose can hope for is B? Nash argues that if N was fairer than M with C available, it is still fairer if C is not available. Fairness is *not* to be affected by the availability or unavailability of unrealistic, pie-in-the-sky alternatives. Not everyone who has considered the problem agrees. See Exercise 7 for a different approach.

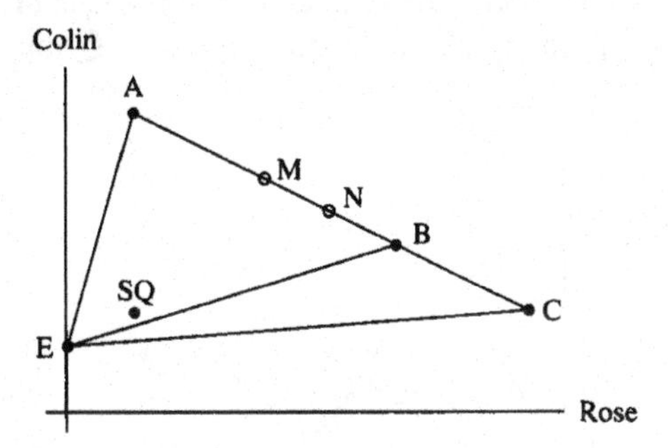

Figure 16.3: What independence of irrelevant alternatives rules out

If we agree to accept Nash's four axioms as conditions which an arbitration scheme should satisfy, then the arbitration scheme we should use is completely determined, for Nash proved the following remarkable theorem:

THEOREM [NASH, 1950b]. There is one and only arbitration scheme which satisfies Axioms 1 through 4. It is this: if SQ $= (x_0, y_0)$, then the arbitrated solution point N is the point (x, y) in the polygon with $x \geqslant x_0$ and $y \geqslant y_0$ which *maximizes the product* $(x - x_0)(y - y_0)$.†

†If the maximum value of this product is positive, the point (x, y) is uniquely determined. If the maximum of the product is zero, it must be because either $y - y_0$ or $x - x_0$ is never positive. If $y - y_0$ is never positive, the solution point is defined to be (x, y_0) where x is the largest value such that this point is in the polygon, and similarly if $x - x_0$ is never positive.

The theorem is surprising first because of the uniqueness statement: this arbitration scheme is the *only* scheme which will satisfy Nash's axioms. Second, it is surprising because it requires us to maximize the *product* of the players' utility gains from the status quo point. That seems mysterious, but notice that it does fit well with Axiom 2. If we multiplied all the x values, say, by a positive constant, the product $(x - x_0)(y - y_0)$ would be multiplied by that constant, and the outcome which maximized the product before would still maximize it after. In fact, it is quite easy to check that this arbitration scheme satisfies all four of the Nash axioms. It is a little harder to prove that it is the only scheme which does, but the proof is quite instructive, and worth following closely.

PROOF. We start with any polygon $\mathcal{Q}$, with status quo point (x_0, y_0). Denote by N the point in $\mathcal{Q}$ with $x \geqslant x_0$, $y \geqslant y_0$ which maximizes the product $(x - x_0)(y - y_0)$. We must show that any arbitration scheme which satisfies Axioms 1–4 must choose N as the solution point for this situation.

Using Axiom 2, we first subtract x_0 from all the x-utilities and y_0 from all the y-utilities, so that the status quo point is at $(0, 0)$. We then multiply the x- and y-utilities by positive constants so that the point N is at $(1, 1)$.† Then by the product-maximizing property of N, the entire polygon $\mathcal{Q}$ must lie on or below

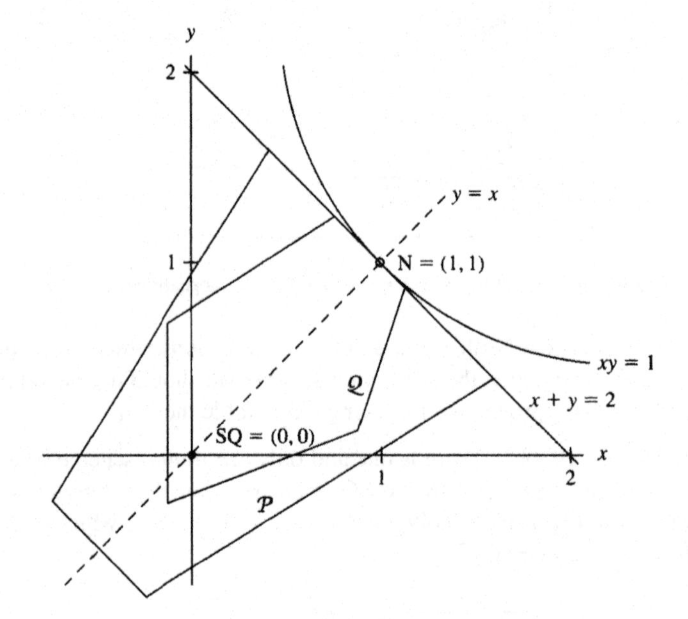

Figure 16.4: The proof of Nash's Theorem

†If the solution point has $x = x_0$ or $y = y_0$ or both, a different argument is needed, but I will leave that to you to supply.

the positive branch of the hyperbola $xy = 1$ (see Figure 16.4). In fact, since the hyperbola is concave upwards and $\mathcal{Q}$ is convex, $\mathcal{Q}$ must lie on or below the tangent line to the hyperbola at $(1, 1)$, which is the line $x + y = 2$. Hence we can enclose $\mathcal{Q}$ in a larger polygon $\mathcal{P}$ which has this tangent line as its northeast boundary, and is symmetric about the line $x = y$. The relationship is pictured in Figure 16.4. Now by Axioms 1 and 3, the solution to the arbitration problem $(\mathcal{P}, (0,0))$ must be N. Then by Axiom 4, the solution to the arbitration problem $(\mathcal{Q}, (0,0))$ must also be N, and we are done!

Notice how each of the axioms comes into play in this argument, and in particular the powerful role of Axiom 4 at the end. In addition to the theoretical elegance of its proof, Nash's solution to the arbitration problem has the advantage of being easy to compute. We will consider two examples.

EXAMPLE 1. Suppose that Rose and Colin must agree on one of the outcomes A = $(0,0)$, B = $(2,0)$, C = $(4,2)$, D = $(1,5)$, or some probability mixture of these outcomes. The numbers for each outcome are, of course, the Rose and Colin utilities for that outcome. If they cannot agree, the outcome will be SQ = $(2, 1)$. As a Nash arbitrator, what would you propose for a fair solution?

The situation is pictured in Figure 16.5a. The Nash point must lie in the negotiation set, which is the line segment from $(2,4)$ to $(4,2)$. It is the point on this line segment which maximizes the product $(x-2)(y-1)$. Since the equation of the line segment is $y = 6 - x$ $(2 \leqslant x \leqslant 4)$, the expression we must maximize is

$$(x-2)(6-x-1) = -x^2 + 7x - 10.$$

For the general quadratic expression

$$-ax^2 + bx + c \qquad (a \geqslant 0)$$

you can use calculus if you know it, or completing the square if you don't, to find that the maximum occurs at $x = b/2a$. Hence for our problem, the maximum is at $x = 7/2$, and $y = 6 - \frac{7}{2} = \frac{5}{2}$. This is $\frac{5}{6}C + \frac{1}{6}D$, which is what you should recommend as the solution.

This problem has the special property that the Pareto optimal boundary is a line segment of slope -1. When that is true, there is an easy geometric way to find the Nash solution point: start at SQ and travel northeast until you hit the boundary, and that point will be the Nash point (see Figure 16.5a). In the exercises you are asked to explain why this is true, and to find out what happens when the negotiation set lies on a line of slope other than -1.

EXAMPLE 2. Suppose that the alternatives are A = $(1,8)$, B = $(6,7)$, C = $(8,6)$, D = $(9,5)$, E = $(10,3)$, F = $(11,-1)$, G = $(-1,-1)$, and the status quo is SQ = $(2,1)$. What is the Nash solution point?

The situation is pictured in Figure 16.5b. The negotiation set is a collection of line segments. To get an idea of where the solution point might lie, we compute

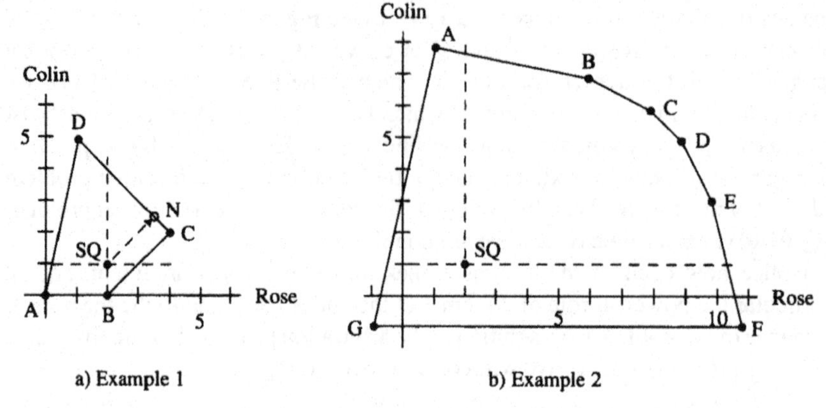

a) Example 1

b) Example 2

Figure 16.5

$(x-2)(y-1)$ for the corner points of the negotiation set:

$$\begin{aligned}
B:&\quad (6-2)(7-1) = 24\\
C:&\quad (8-2)(6-1) = 30\\
D:&\quad (9-2)(5-1) = 28\\
E:&\quad (10-2)(3-1) = 16.
\end{aligned}$$

Since C has the largest product, the solution point will lie on one of the two line segments BC or CD. The line containing BC has equation $y = 10 - \frac{1}{2}x$, so we must maximize

$$(x-2)(10 - \tfrac{1}{2}x - 1) = -\tfrac{1}{2}x^2 + 10x - 18.$$

The maximum occurs at $x = 10$, which is beyond the C end of the line segment BC. Hence the maximum on this line segment is at the endpoint C. Similarly, you can check (Exercise 3) that the maximum of $(x-2)(y-1)$ on the line segment CD occurs at the endpoint C. Hence C is the Nash solution point.

With Nash's solution to the general arbitration problem in hand, we can consider our original problem of finding a fair arbitrated solution to a non-zero-sum game. The payoff polygon for the game gives us the polygon for an arbitration problem. To find the status quo point, we have to ask what will happen if arbitration fails. One possibility is to admit that we *don't know* what will happen, except that each player can assure at least his security level. We thus might take as the status quo the point whose coordinates are the security levels of Rose and Colin. For Game 16.1 we would take $SQ = (3\frac{1}{3}, 6)$. The Nash arbitrated solution then turns out to be $(5\frac{2}{3}, 7\frac{1}{6})$, or $\frac{13}{18}BA + \frac{5}{18}AB$ (Exercise 1). We would recommend to the players that they agree to play BA with probability $\frac{13}{18}$, AB with probability $\frac{15}{18}$. If they were playing the game many times, they could play BA $\frac{13}{18}$ of the time, AB $\frac{5}{18}$ of the time.

Security levels have the advantage of being easy to compute, but we might wonder about using them for a status quo point. Nash himself, when he considered the arbitration problem for non-zero-sum games, proposed a different idea [Nash, 1953]. He argued that in an arbitration situation for a game, players might well try to influence the outcome by using *threat strategies*: "If negotiations break down, I will play the following strategy, which will be bad for you." The proper status quo point would then be the outcome determined by these threat strategies. Knowing this, each player would choose his threat strategy to obtain a status quo point as favorable to him as possible, knowing that his opponent was doing the same thing. In other words, the choice of threat strategies is itself a game! However, this game is zero-sum, and Nash showed that it always has a solution, which gives an *optimal threat strategy* for each player. Nash proposed that the status quo point for the arbitration should be the outcome determined by these optimal threat strategies.

In complicated cases, Nash optimal threat strategies may be tricky to compute. There is one case, though, where the computation is very easy: when the negotiation set consists of just one line segment whose slope is -1. In this case, we have seen that the Nash solution point can be obtained from the status quo point by moving northeast along a line of slope $+1$. Thus the *difference* between Rose's payoff and Colin's payoff at SQ will be preserved in the arbitrated solution. Rose would like this difference to be as large as possible, while Colin would like it to be as small is possible. Thus Rose and Colin would find their optimal threat strategies by playing a zero-sum *game of differences* derived from the original non-zero-sum game. For example, consider

		Colin	
		A	B
Rose	A	(2, 12)	(10, 10)
	B	(4, 16)	(0, 0)

Game 16.2

The negotiation set is the line segment from (4,16) to (10,10), which has slope -1. The game of differences is

		Colin	
		A	B
Rose	A	-10	0
	B	-12	0

which has a saddle point at AA. Hence strategy A is the optimal threat strategy for both players, and Nash would take the status quo point for an arbitrated solution to Game 16.2 to be SQ = (2, 12). Of course, in other games the optimal threat strategies could turn out to be mixed strategies.

If the negotiation set of the game polygon is any single line segment, which must have negative slope, the slope can be made to be -1 by multiplying the utilities of one player by an appropriate constant, and then the procedure of the last paragraph can be applied to find optimal threat strategies. For example, Game 16.1 had its negotiation set on a line of slope $-1/2$. If we multiply Colin's payoffs by two, the slope becomes -1, and in fact we get exactly Game 16.2. Thus the optimal threat strategies are A and A, and the Nash status quo would be at AA $= (2, 6)$. With this status quo, the Nash arbitrated solution turns out to be $(5, 7\frac{1}{2}) = \frac{5}{6}$BA $+ \frac{1}{6}$AB (Exercise 1). This is more favorable to Colin than the solution we got when we used security levels to get the status quo point. We noted earlier that Colin has a strategic advantage in this game, which should be reflected in any fair arbitrated solution. Part of this strategic advantage is reflected in Colin's higher security level, but more of it is reflected in Colin's stronger ability to threaten.

Our conclusion is that if you use the Nash arbitration scheme to handle the general arbitration problem, your arbitrated solutions for non-zero-sum games will depend on how you determine the status quo point. This in turn has to do with how you think strategic advantage should be taken into account. Working with further examples in Exercises 5 and 6 might help you begin to make up your mind between the security-level method and Nash's method of optimal threat strategies. There are further discussions and examples in [Rapoport, 1970a], [Luce and Raiffa, 1957], [Owen, 1982], [Bacharach, 1977], and [Jones, 1980].

In the next chapter, we will consider how the general Nash arbitration scheme might be used to arbitrate a labor dispute.

Exercises for Chapter 16

1. For Game 16.1 verify the following claims:
   a) Rose's security level is $3\frac{1}{3}$ and Colin's security level is 6.
   b) With status quo point $(3\frac{1}{3}, 6)$, the Nash solution is $(5\frac{2}{3}, 7\frac{1}{6})$. This can be interpreted as $\frac{13}{18}$BA $+ \frac{5}{18}$AB.
   c) With the threat status quo point $(2, 6)$, the Nash solution is $(5, 7\frac{1}{2})$. This can be interpreted as $\frac{5}{6}$BA $+ \frac{1}{6}$AB.

2. a) Explain why, when the Pareto optimal set of a polygon is a single line segment of slope -1, the Nash solution can be found by starting at the status quo point and following a line of slope $+1$ to the negotiation set. You might think in terms of the axioms, and look at the proof of Nash's theorem.
   b) Find a similar geometric method for locating the Nash solution when the negotiation set is a single line segment of slope $-m$. Why does your method work?
   c) Check your method by solving the arbitration problem A $= (3, 7)$, B $= (6, 1)$, C $= (0, 0)$, SQ $= (3, 1)$.

3. Verify the claim in the solution to Example 2, that the maximum value of $(x - 2)(y - 1)$ on the line containing CD lies to the left of C.

4. Solve the following arbitration problems using the Nash arbitration scheme:
   a) A = (5, 19), B = (12, 14), C = (14, 12), D = (16, 10), E = (20, 0), F = (−5, −7), G = (0, −10), SQ = (0, −7).
   b) A = (6, 3), B = (4, 7), C = (2, 8), D = (3, 6), E = (−2, 3), F = (2, −1), SQ = (2, 1).

   Express your solutions as probability mixtures of the outcomes.

5. Compare the arbitrated solutions using the security level SQ vs. the optimal threat strategy SQ for the following games:

   a)

		Colin A	Colin B
Rose	A	(2, 7)	(7, 2)
	B	(0, 1)	(1, 0)

   b)

		Colin A	Colin B
Rose	A	(0, −10)	(10, 0)
	B	(0, 10)	(−10, 0)

   Which method seems to you fairest for these games?

6. R. B. Braithwaite [1955] considered the problem of arbitrating the following game between Luke and Matthew:

		Matthew A	Matthew B
Luke	A	(0, 1)	(6, 2)
	B	(3, 9)	(1, 0)

   and proposes a method which gives the solution $\frac{17}{43}$AB + $\frac{26}{43}$BA = (4.19, 6.23). Compare this to the Nash arbitration solution using
   a) the security level status quo
   b) the optimal threat strategy status quo.

7. Kalai and Smorodinsky [1975] considered the following related arbitration problems, both with status quo point at (0,0):

   $\mathcal{Q}$ has vertices at (0, 0), (20, 0), (0, 20), (15, 15).

   $\mathcal{P}$ has vertices at (0, 0), (20, 0), (0, 20), (20, 14).

   a) Show that the Nash solution for $\mathcal{Q}$ is (15, 15), and the Nash solution for $\mathcal{P}$ is (20, 14).
   b) Draw the polygons to see that $\mathcal{P}$ completely contains $\mathcal{Q}$. Kalai and Smorodinsky argue that these examples show that the Nash arbitration scheme is unfair, since if new outcomes become available which would allow Colin to get more, for any given payoff to Rose, a fair arbitration scheme should *not* give Colin less. They go on to propose an alternative arbitration scheme which does not have this problem. (It does not, of course, satisfy all of Nash's axioms.)

17. Application to Business: Management-Labor Arbitration

The management of a factory is negotiating a new contract with the union representing its workers. The union has demanded new benefits for its members: a one dollar per hour across-the-board raise, and increased pension benefits. In turn, management has demanded concessions from the union. Management would like to eliminate the 10:00 a.m. coffee break, which has proven to be excessively costly as workers straggle slowly back to the assembly line, and to automate one of the assembly line checkpoints. The union opposes both demands, especially the automation, which would eliminate union jobs. The dispute has not been resolved, but both sides are willing to try arbitration before resorting to stronger measures. You have been called in as an arbitrator. Could you find a solution which would be fair, and please both sides?

In this chapter, following [Allen, 1956], we will consider how the Nash arbitration scheme might be used to solve such a problem. The first step would be to sit down with management and labor separately and determine their utilities for the various proposals under consideration:

A: automation of the checkpoint

C: elimination of the coffee break

R: a one dollar per hour raise for workers

P: increased pension benefits

SQ: the status quo

For convenience, we can choose the utility scales so that SQ has utility zero for both sides. We then expect management to have positive utilities for A and C and negative utilities for R and P, while labor's utilities will have the opposite signs. We may well find, however, that preferences are not strictly opposed, so that we are not in a zero-sum situation. In particular, there may be "trades" which are better than SQ for both sides. We would like to find such a trade, which is, moreover, fair to both sides.

We first ask management to rank the alternatives. Suppose they value A and C equally, and would rather give in on the pension benefits than on the raise, so that we have an ordinal utility ranking A = C, SQ, P, R. We then ask further questions to determine cardinal utility. These might take the form of lottery questions as in Chapter 9, but there might be other more natural ways in which questions could be put. For instance, we might determine that management would be indifferent between granting the pension benefits and giving a raise of $.67 per hour, and that management would be willing to trade the full one dollar raise and half of the pension benefits for elimination of the coffee break. That would give cardinal utilities

R	P	SQ	A, C	
-3	-2	0	4	Management's utilities

We then ask similar questions of the union, say obtaining

A	C	SQ	P	R	
-2	-1	0	2	3	Labor's utilities

Of course, the scales are arbitrary up to linear functions, and that will not affect our further analysis since the Nash arbitration scheme is invariant under linear functions.

We now use this information to evaluate management and labor utilities for all possible "trades." The simplest case is when the utilities are additive, in that if labor values getting P at 2 and getting R at 3, it will value getting both P and R at $2 + 3 = 5$. We will assume that this is true, although there are certainly cases where benefits or concessions would not behave like this. For example, getting P might be more valuable (or less valuable) if labor gets R, than if it doesn't. Those cases could be handled by asking separate utility questions about trades with significant linkages. Here are the utilities of trades in our example:

		Labor concedes			
		Nothing	C	A	CA
Management concedes	Nothing	$(0,0)$	$(4,-1)$	$(4,-2)$	$(8,-3)$
	P	$(-2,2)$	$(2,1)$	$(2,0)$	$(6,-1)$
	R	$(-3,3)$	$(1,2)$	$(1,1)$	$(5,0)$
	PR	$(-5,5)$	$(-1,4)$	$(-1,3)$	$(3,2)$

We draw the payoff polygon for these outcomes, shown in Figure 17.1. We then determine the Nash arbitration point, using $(0,0)$ as the status quo point. The solution in this case is $(3,2)$, corresponding to the trade PRCA where each side grants both of the demands of the other side. Both sides are better off than at the status quo, and the Nash axioms justify the result as fair.

The process seems appealing, doesn't it? Of course, there are problems with it, some of which might have occurred to you. We have already mentioned that non-linearities of utility might complicate the analysis. It is also true (see Exercise 2) that increasing the number of management and labor issues increases the complexity of the problem dramatically, to the point where a computer would become helpful. We will close by considering four problems of a more fundamental nature.

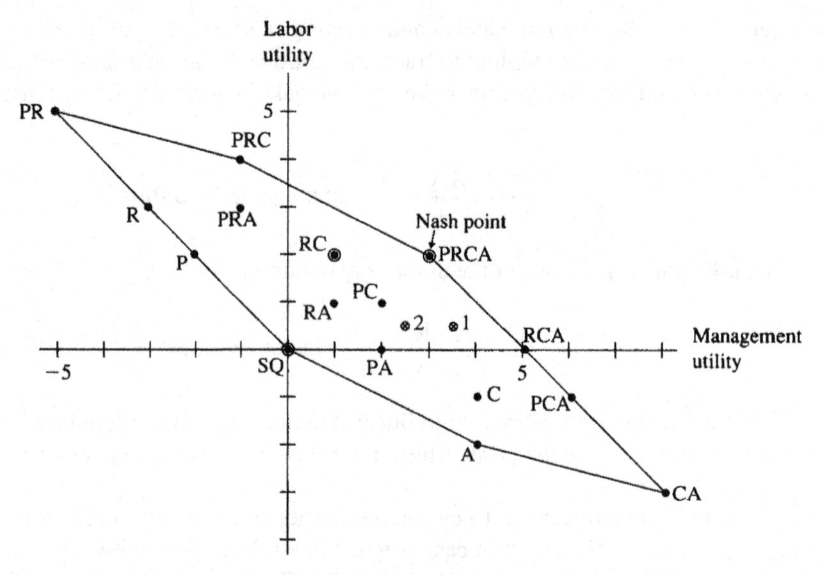

Figure 17.1: Payoff polygon for labor-management arbitration.

1) What if the Nash point is a mixed outcome? For example, suppose that the Nash solution in the example had been at $(2, 2\frac{1}{2}) = \frac{1}{4}\text{PRC} + \frac{3}{4}\text{PRCA}$. We would recommend giving the pension benefits, the raise, and eliminating the coffee break, but how could we give $\frac{3}{4}$ of the automation? One answer might be to grant the automation but require that $\frac{1}{4}$ of the displaced workers be guaranteed other jobs. Some demands are of a "divisible" nature, like raises and pension benefits. Others may be harder to divide: half a coffee break would not satisfy either side. However, perhaps we could eliminate the coffee break but lengthen the lunch hour. Problems of this kind can often be dealt with by exercising a little creativity.
2) What do we do if there are no outcomes which are Pareto superior to the status quo? Well, we could just recommend the status quo. However, a more creative solution would be to try to enlarge the collection of proposals under consideration. The more alternatives we have to consider, the more chance there will be of having enough alternatives which the sides value differently to arrange mutually profitable trades. [Allen, 1956] recommends that the very first step of the arbitrator should be to hold "brainstorming" sessions with labor and management to generate additional proposals. Management might offer to provide a college scholarship for union children (getting the resulting tax write-off). Labor might offer to modify a grievance procedure which has been bothering management. These new ideas would then go into the process along with the original issues. This idea of enlarging

the scope of the bargaining to increase chances of mutually profitable agreements has been strongly endorsed by many analysts, for instance by Fisher and Ury in a best-selling book on negotiation techniques [1981].

3) Is the "present situation" the appropriate status quo point for the Nash procedure? To assume it is, is to assume that negotiations are friendly efforts in good faith. In fact, labor-management negotiations often occur in an atmosphere of threats, with talk of strikes and lockouts. Such actions would be costly to both sides, but they can be thought of as attempts to alter the Nash status quo point in favor of one side. An arbitrator might have to consider whether or how to react to such attempts. It would not be easy.

4) Wouldn't it be natural for both sides to give false information about their utilities, thus rendering the whole analysis useless? This objection is fundamentally important, and we should consider it carefully. If one or both sides knew that it could obtain a more favorable result by lying about its utilities, the process would indeed be useless in practical situations. However, the situation is more subtle than that: it is not at all certain that lying will be profitable, and whether it is or not depends on what the other side is doing.

To show the subtleties involved, let's consider some possible effects of misrepresenting utilities in our example. Suppose management is considering lying about its utilities. What kind of lie might be beneficial? One promising technique might be to give correct relative scaling for the positive utility items, and correct relative scaling for the negative utility items, but to misrepresent the relationship between the two by answering "trading" questions falsely. Management reports that it would require many concessions from the other side before it would give in to any of their demands. This kind of misrepresentation is often what we have in mind when we talk about "hard-nosed bargaining." For instance, management might double all the negative numbers, and report false utilities

R	P	SO	A, C
−6	−4	0	4

Management's false utilities

To see what effect this misrepresentation would have on the arbitrated outcome, we redo our analysis using management's false utilities:

		Labor concedes			
		Nothing	C	A	CA
	Nothing	(0, 0)	(4, −1)	(4, −2)	(8, −3)
Management	P	(−4, 2)	(0, 1)	(0, 0)	(4, −1)
concedes	R	(−6, 3)	(−2, 2)	(−2, 1)	(2, 0)
	PR	(−10, 5)	(−6, 4)	(−6, 3)	(−2, 2)

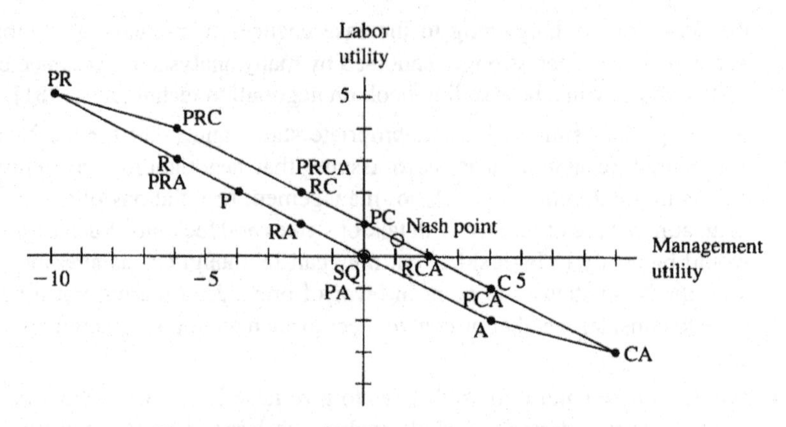

Figure 17.2: Payoff polygon when management lies.

The payoff polygon is shown in Figure 17.2. The change from Figure 17.1 is that management's misrepresentation has moved some points to the left, making the area of mutually profitable agreements smaller. The new Nash point is at $(1, \frac{1}{2})$, which can be interpreted in several different ways. It could, for example, be implemented as $\frac{1}{2}$PC + $\frac{1}{2}$RCA. If we look back at the honest utilities, this point is $(3\frac{1}{2}, \frac{1}{2})$, shown as point #1 in Figure 17.1. This point is not Pareto optimal, but it is better for management than the honest Nash point of $(3, 2)$. Management has gained by lying.

On the other hand, the dishonest Nash point $(1, \frac{1}{2})$ might also be implemented as $\frac{3}{4}$PC + $\frac{1}{4}$C. The honest utilities of this implementation are $(2\frac{1}{2}, \frac{1}{2})$, shown as point #2 in Figure 17.1. This is worse for management than the honest Nash point. In other words, lying might help management, but it could also hurt them.

In Exercise 3, you can see that if labor lies by doubling its negative utilities, while management tells the truth, the resulting solution could be implemented as RC, which is point $(1, 2)$ in honest utilities. In this case, labor's misrepresentation has no effect on labor's payoff, although it hurts management and destroys Pareto optimality. Finally, in Exercise 4 you can check that if both labor and management choose to lie, the Nash arbitration scheme could recommend the status quo, and both sides would lose the chance of making a mutually profitable agreement.

The general situation is that lying by one side tends to produce Pareto inferior outcomes, but whether it will help or hurt the liar is a subtle matter. Lying by both sides will generally hurt both sides. There may be profit to be made by misrepresenting utilities, but it is not easy or certain profit.

I am afraid that I do not know of a case in which the Nash arbitration scheme has been used to arbitrate a real management-labor dispute. If you do, I would like to hear! It seems to me worth trying. One of my students has applied Nash's scheme to a labor dispute which had already been resolved by other means [Polgreen, 1992]. The most interesting part of the process was getting representatives of management and labor to answer cardinal utility questions. For a discussion of possibilities and problems with using the Nash scheme, see Chapter 16 of [Raiffa, 1982]. Other game-theoretic approaches to arbitration and bargaining are given in [Brams, 1990].

Exercises for Chapter 17

1. Verify that the Nash solution point for the example of Figure 17.1 is $(3, 2)$.
2. If management had m proposals and labor had l, how many trades would be possible? At what point would you want to use a computer to help you solve the problem?
3. Suppose in our example that management gave true utilities but labor lied by doubling all of its negative utilities. Show that one possible solution to the resulting arbitration problem would be RC.
4. Show that if both management and labor lie by doubling their negative utilities, then SQ is a Nash arbitrated outcome.

18. Application to Economics: The Duopoly Problem

A *duopoly* is a situation in which two companies control the market for a certain commodity. The duopoly problem is to decide how the companies in a duopoly situation should adjust their production to maximize their profits. In this chapter we will use a simple example to compare four different "solutions" to the duopoly problem. Some of the solutions involve calculus—at least knowing how to differentiate polynomials and the fact that a maximum value of a function occurs at a point where its derivative is zero. If you know calculus, you can calculate these solutions along with me; if not, you'll have to take my word for them. This chapter follows [Mayberry, Nash, and Shubik, 1953] and some additional calculational details can be found there.

Here are the variables we will be concerned with:

q_i = the number of items (in thousands) produced by company i ($i = 1, 2$).

AC_i = the average production cost per item for company i.

$TC_i = q_i \cdot AC_i$ = the total production cost (in \$1000) for company i.

$MC_i = \dfrac{d(TC_i)}{dq_i}$ = the *marginal cost*, the cost per item of raising production slightly, for company i.

p = the price per item at which the commodity can be sold.

$P_i = q_i \cdot p - TC_i$ = the profit (in \$1000) of company i.

The average cost per item varies with the quantity produced, and in our example we will posit the following functions:

$$AC_1 = 64 - 4q_1 + q_1^2 \qquad AC_2 = 80 - 4q_2 + q_2^2.$$

These average cost functions are graphed in Figure 18.1. For both companies the average cost per item at first decreases due to economies of scale. It reaches a minimum at $q = 2$ and then begins to increase as the company has to invest more capital and hire more labor to increase production further. Company 1 is more efficient than company 2, and can produce the commodity at \$16 less per item at any level of production.

We then calculate, taking derivatives for the marginal costs:

$$TC_1 = 64q_1 - 4q_1^2 + q_1^3 \qquad TC_2 = 80q_2 - 4q_2^2 + q_2^3$$
$$MC_1 = 64 - 8q_1 + 3q_1^2 \qquad MC_2 = 80 - 8q_2 + 3q_2^2$$

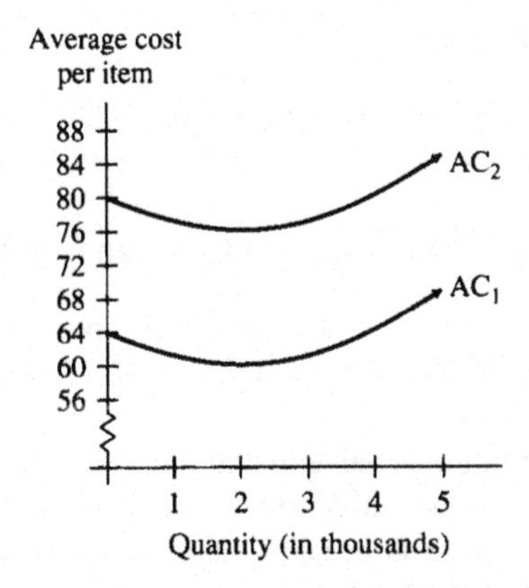

Figure 18.1: Average cost functions for the duopoly example.

We will also assume a relationship between the total quantity produced and the price per item at which the commodity can be sold:

$$p = 160 - 8(q_1 + q_2).$$

If the items are very rare they can be sold for \$160 each, but as production increases and the market becomes glutted, the price must be lowered to sell all of the items produced.

Finally, we can compute

$$\begin{aligned} P_1 &= q_1(160 - 8q_1 - 8q_2) - (64q_1 - 4q_1^2 + q_1^3) \\ &= 96q_1 - 4q_1^2 - q_1^3 - 8q_1q_2 \end{aligned}$$

and similarly

$$P_2 = 80q_2 - 4q_2^2 - q_2^3 - 8q_1q_2.$$

Each company's goal is to choose its production level q_i so as to maximize its profit P_i, and our problem is to understand how the companies might go about doing this. One complication is that each company's profit is determined not only by its own choice of production level, but also by the other company's choice, through the mixed term $-8q_1q_2$. We are in a game situation. Following [Mayberry, Nash, and Shubik, 1953] we will consider four different ways companies might approach this problem, in order of increasing strategic sophistication.

First, consider a classical economic approach which is completely non-strategic. Each company starts off producing a fairly small amount, and slowly increases its production as long as the cost of producing additional items is less than the price at which items can be sold. In other words, each company increases its production to the point where its marginal cost is equal to the selling price of the commodity. This point is called the *efficient point*, and we find it by solving the equations $MC_1 = p = MC_2$, i.e.

$$64 - 8q_1 + 3q_1^2 = 160 - 8q_1 - 8q_2 = 80 - 8q_2 + 3q_2^2,$$

which reduce to

$$8q_2 = 96 - 3q_1^2 \qquad 8q_1 = 80 - 3q_2^2,$$

whose solution is $q_1 = 4.69$, $q_2 = 3.76$. If the companies produce these quantities, the selling price is \$92 per item, and the profits are $P_1 = 118$, $P_2 = 50$ (see Exercise 1 and Table 18.1).

The efficient point solution is naive in that it ignores the fact that the quantities which both companies produce affect the selling price. If they produced less the price would be higher, and it is possible that this might yield higher profits. It is time to examine the situation as a two-person game between Company 1 and Company 2:

		q_2				
		2	2.5	3	3.5	4
	2	136, 104	128, 119	120, 129	112, 132	104, 128
	2.5	159, 96	149, 109	139, 117	129, 118	119, 112
	3	177, 88	165, 99	153, 105	141, 104	129, 96
q_1	3.5	188, 80	174, 89	160, 93	146, 90	132, 80
	4	192, 72	176, 79	160, 81	144, 76	128, 64
	4.5	188, 64	170, 69	152, 69	134, 62	116, 48
	5	175, 56	155, 59	135, 57	115, 48	95, 32

(Profits in \$1000)

Due to limitations of space, I've only shown the payoffs for values of q_1 and q_2 at intervals of 0.5. The game is clearly non-constant-sum. The efficient point is in the lower right part of the table. Notice that that both companies can indeed raise their profits by restricting their production.

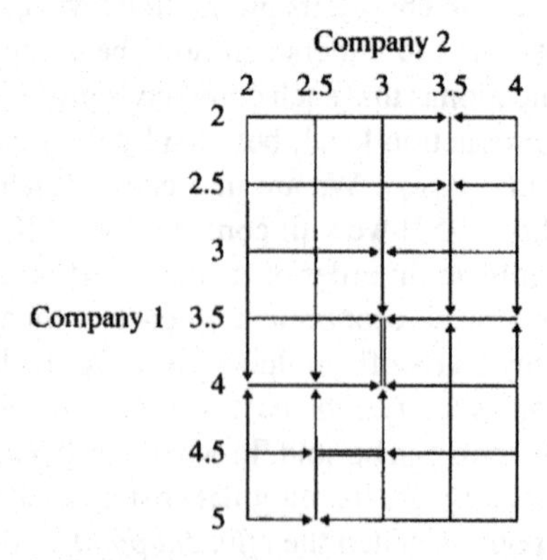

Figure 18.2: Movement diagram for the duopoly game.

To see what might happen if the companies do this, consider the movement diagram for this game in Figure 18.2. The game has a unique Nash equilibrium somewhere near $q_1 = 3.75, q_2 = 3$. To find it exactly, note that it is an outcome where neither player can raise his own profits P_i by changing q_i. Hence we must have

$$\frac{\partial P_1}{\partial q_1} = 96 - 8q_1 - 3q_1^2 - 8q_2 = 0$$

$$\frac{\partial P_2}{\partial q_2} = 80 - 8q_2 - 3q_2^2 - 8q_1 = 0.$$

The solution to this set of equations is $q_1 = 3.75$, $q_2 = 2.96$. The price is \$106, and the profits are $P_1 = 162$, $P_2 = 87$ (see Exercise 2 and Table 18.1). This solution to the duopoly problem predates game theory, and in fact comes from the work of Cournot in the 1830's. It is known in economics as the Cournot equilibrium.

The Nash-Cournot equilibrium in this game is not Pareto optimal, as we see when we plot payoff pairs in Figure 18.3. Both companies could do better if they could cooperate. If we allow cooperation and compute the Nash arbitrated solution (with the threat-point status quo—see Exercise 3), we find it to be at $q_1 = 3.30$, $q_2 = 2.40$, with a price of \$114 and profits $P_1 = 174$, $P_2 = 91$. This is an improvement for both companies. Notice that this improvement is achieved

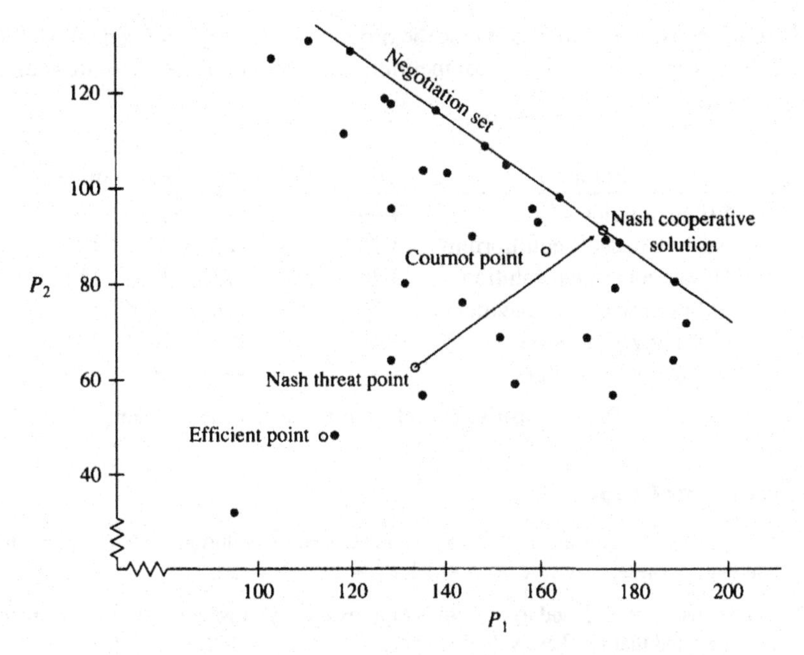

Figure 18.3: Payoff pairs for the dupoly game.

by restricting total production from 6.7 at the non-cooperative equilibrium to 5.7 at the cooperative solution, with a resulting price rise from \$106 to \$114 (see Table 18.1). Because cooperation in a duopoly situation will in general restrict production, raise prices and harm the consumer, it is often called "collusion" and prohibited by law.

Finally, if it is possible for one company to make *side-payments* to the other company, the companies can do even better at $q_1 = 3.66$, $q_2 = 1.66$, where the total profit is maximized at 269 ($P_1 = 200$, $P_2 = 69$). If Company 1 made a side-payment of 24 to Company 2 so that the profits were split 176 to 93, both would be even better off than at the Nash cooperative solution. Of course, this improvement involves a further restriction of production and a raising of the price to \$117. Side-payments between companies are quite often prohibited by law.

Table 18.1 compares all of the duopoly solutions we have discussed, and also the results if either company had a monopoly and maximized profits (Exercise 4). From the companies' point of view, the solutions can be ranked from most desirable to least desirable:

- monopoly
- cooperation with side-payments
- cooperation without sidepayments
- the non-cooperative game equilibrium
- the efficient point.

From the consumer's point of view, the ordering is directly reversed. Table 18.1 clearly illustrates the role of competition and anti-collusion laws in keeping prices low.

Solution	q_1	q_2	P_1	P_2	price
Efficient point	4.69	3.76	118	50	92
Nash (Cournot) equilibrium	3.75	2.96	162	87	106
Nash cooperative solution	3.30	2.40	173	91	114
Solution with side-payments	3.66	1.66	176	93	117
Company 1 monopoly	4.48	—	260	—	124
Company 2 monopoly	—	4.00	—	192	128

Table 18.1: Comparison of solutions to the duopoly problem.

Exercises for Chapter 18

1. Verify that $q_1 = 4.69$ and $q_2 = 3.76$ is (approximately) a solution to $MC_1 = p = MC_2$, and that for these values we get $p = 92$, $P_1 = 118$, $P_2 = 50$.

2. Verify that $q_1 = 3.75$ and $q_2 = 2.96$ is (approximately) a solution to $\partial P_1/\partial q_1 = 0 = \partial P_2/\partial q_2$, and that for these values we get $p = 106$, $P_1 = 162$, $P_2 = 87$.

3. a) The Nash threat point for this duopoly game turns out to be about (134, 62), as shown on Figure 18.3. Notice that the negotiation set is on a line of slope about $-\frac{3}{4}$. Use these facts to show that the Nash arbitrated solution is about $(P_1, P_2) = (173, 91)$.
   b) Verify that $q_1 = 3.3$ and $q_2 = 2.4$ gives these values for P_1 and P_2.

4. (If you know calculus)
   a) If company 1 had a monopoly, we would know that $q_2 = 0$, so that

$$P_1 = 96q_1 - 4q_1^2 - q_1^3.$$

   Show that P_1 is maximized at $q_1 = 4.48$, and then $p = 124$ and $P_1 = 260$.
   b) Show that if company 2 had a monopoly and maximized profits, the result would be $q_2 = 4, p = 128, P_2 = 192$.

Part III

N-Person Games

19. An Introduction to *N*-Person Games

Until now, we have dealt only with games played between two players. In our modern interconnected world, such games are rare. Most important economic, social, and political games involve more than two players. We will now turn our attention to n-person games, where n is assumed to be at least three. We will find that with three or more players, new and interesting difficulties appear.

To begin our analysis, let us consider the simplest possible case, a three-person $2 \times 2 \times 2$ zero-sum game. Game 19.1 is an example.

Larry A

		Colin A	Colin B
Rose	A	$(1, 1, -2)$	$(-4, 3, 1)$
	B	$(2, -4, 2)$	$(-5, -5, 10)$

Larry B

		Colin A	Colin B
Rose	A	$(3, -2, -1)$	$(-6, -6, 12)$
	B	$(2, 2, -4)$	$(-2, 3, -1)$

Game 19.1

The three players are Rose, Colin and Larry (Larry chooses the "layer"). Each outcome is a triple of numbers giving the payoff to these three players in that order. There are $2 \times 2 \times 2 = 8$ possible outcomes, which could be positioned in a three-dimensional array. For convenience on a two-dimensional page, the outcomes for Larry A and Larry B are given in two separate two-dimensional tables. The game is zero-sum since the three payoffs in each outcome add to zero.

As in two-player games we can search for pure strategy equilibria by drawing a movement diagram. Two possible forms are shown in Figure 19.1. The three-dimensional diagram is clear, but a little hard to draw well. In the two-dimensional diagram, arrows point out of Larry A and into Larry B when Larry prefers his payoff for B to his payoff for A. For example, if Rose and Colin both play A, Larry prefers his payoff -1 at AAB to his payoff -2 at AAA. Hence at Rose

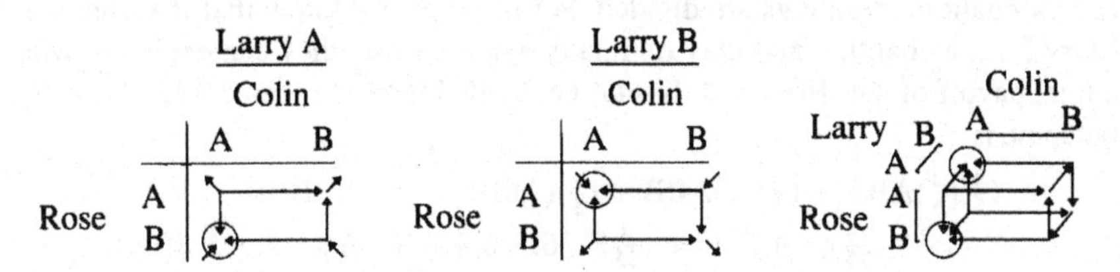

Figure 19.1: Two forms of movement diagram for Game 19.1.

A–Colin A the arrow points out of the Larry A diagram and into the Larry B diagram.

The movement diagram indicates that none of the three players has a dominant strategy, and that there are two pure strategy equilibria, at BAA = $(2, -4, 2)$ and AAB = $(3, -2, -1)$. Notice that these equilibria are not equivalent and not interchangeable. In fact, Rose and Colin would prefer the equilibrium at AAB, while Larry would prefer the equilibrium at BAA. If Rose plays A to try for her favorite equilibrium and Larry plays A to try for his favorite, the result will be AAA, which is *not* an equilibrium. This example shows that the major difficulties of two-person non-zero sum games appear even in zero-sum games as soon as there are three or more players. We will not be able to find a simple solution theory.

A further difficulty appears if we allow players to communicate: there may be a strong temptation for two of the players to form a *coalition* against the third player. Let us investigate carefully how this might work in Game 19.1. Suppose Colin and Larry form a coalition and agree to coordinate their play against Rose. We can represent the result as a two-player game of Rose against the combined player Colin-and-Larry. Since this game is zero sum, we can just give the payoffs to Rose, obtaining the 2×4 game in Figure 19.2a.

The solution for this game has Rose playing $\frac{3}{5}$A, $\frac{2}{5}$B and receiving an expected payoff of -4.4. Since this is the best Rose can do in the worst possible situation, when Colin and Larry gang up and play against Rose, it makes sense to call this strategy Rose's *prudential strategy*, and the payoff -4.4 Rose's *security level*. The analysis also tells us what Colin and Larry should do if they decide to form a coalition with the goal of winning as much as possible from Rose. Colin should always play B, and Larry should play $\frac{4}{5}$A, $\frac{1}{5}$B. The coalition will win an expected payoff of 4.4 from Rose.

It would also be possible for Rose and Larry to form a coalition against Colin, or for Rose and Colin to form a coalition against Larry. These possibilities are shown in Figure 19.2b and c. I would suggest checking each of the 2×4 games to be sure that you understand where the payoffs come from. We see that Colin's prudential strategy is Colin A, yielding a security level of -4. Larry's prudential strategy is $\frac{3}{7}$A, $\frac{4}{7}$B with a security level of -1.43.

It is clear from this analysis is that if coalitions are possible, each of the players would like to be in one. Being left out is costly. But which of the three possible coalitions would form? One approach to answering this question might be to look at how coalition winnings are divided. For instance, we know that if Colin and Larry form a coalition and play optimally against Rose, they can expect to win a total payoff of 4.4. How will this 4.4 be divided between them? The expected outcome is

$$\begin{aligned}
&(\tfrac{3}{5})(\tfrac{4}{5})\text{ABA} + (\tfrac{3}{5})(\tfrac{1}{5})\text{ABB} + (\tfrac{2}{5})(\tfrac{4}{5})\text{BBA} + (\tfrac{2}{5})(\tfrac{1}{5})\text{BBB} \\
&\quad = (\tfrac{12}{25})(-4, 3, 1) + (\tfrac{3}{25})(-6, -6, 12) + (\tfrac{8}{25})(-5, -5, 10) \\
&\qquad + (\tfrac{2}{25})(-2, 3, -1) \\
&\quad = (-4.40, -0.64, 5.04).
\end{aligned}$$

a)

Colin-and-Larry

		AA	BA	AB	BB	Rose optimal:
Rose	A	1	−4	3	−6	$\frac{3}{5}$
	B	2	−5	2	−2	$\frac{2}{5}$
Colin-and-Larry	optimal:		$\frac{4}{5}$		$\frac{1}{5}$	Value = −4.4

b)

Rose-and-Larry

		AA	BA	AB	BB	Colin optimal:
Colin	A	1	−4	−2	2	1
	B	3	−5	−6	3	0
Rose-and-Larry	optimal:		1	0		Value = −4

c)

Rose-and-Colin

		AA	BA	AB	BB	Larry optimal:
Larry	A	−2	2	1	10	$\frac{3}{7}$
	B	−1	−4	12	−1	$\frac{4}{7}$
Rose-and-Colin	optimal:	$\frac{6}{7}$	$\frac{1}{7}$			Value = −1.43

Figure 19.2: Coalitions in Game 19.1.

This calculation is enlightening. It is Larry who does well in this coalition! On the other hand, notice that Colin, although not very well off, is still not nearly as badly off as if he had been left out, with Rose and Larry forming a coalition against him.

The results for the other possible coalitions are

Colin vs. Rose-and-Larry:	$(2.00, -4.00, 2.00)$
Larry vs. Rose-and-Colin:	$(2.12, -0.69, -1.43)$.

These figures might be used to say which coalition should form, as follows. For each player, find that player's preferred coalition partner. For instance Rose would prefer Colin as a coalition partner, since she wins 2.12 in a coalition with Colin, compared to only 2.00 in a coalition with Larry. Similarly, Colin's preferred coalition partner is Larry, and Larry's preferred coalition partner is Colin. Since Larry and Colin prefer each other, we might expect that the coalition Colin-and-Larry would form. Unfortunately, in other three-person games it can happen that no pair of players prefer each other (see Exercise 3), and it is not clear what we would expect in that case.

Von Neumann and Morgenstern [1944] made an additional assumption to deal with coalitions in n-person games. They assumed that *sidepayments* are possible between players. Thus in Game 19.1 Rose could offer Colin a sidepayment of 0.1 to join in a coalition with her, making the effective payoffs for the Rose-Colin coalition $(2.02, -0.59, -1.43)$. Now this coalition would be more attractive to Colin than the Colin-Larry coalition. Of course, there would be nothing to stop

Larry from also bidding for Colin's support, or Rose from offering a sidepayment to Larry.

The assumption that sidepayments are possible is very strong. First of all, it assumes that utility is *transferable* between players. Second, it assumes that utility transferred between players is of *comparable* value to those players—that 0.1 unit to Rose means the same as 0.1 unit to Colin. In Chapter 9 we saw that the general von Neumann-Morgenstern theory of cardinal utility does not justify interpersonal comparisons of utility. Hence if we are to allow sidepayments from one player to another, it must be in some more structured environment where there is a medium of exchange which it is reasonable to assume is valued equally by all players. That is not completely far-fetched. In our daily economic transactions we operate in such an environment, where the medium of exchange is money. Of course if we think about it, we realize that money is probably not valued equally by a millionaire and a pauper, or by a philosopher and a miser, but in many situations it does seem reasonable to assume that it preserves its value when transferred between people.

There is a theory of n-person games without sidepayments, to which one introduction is [Aumann, 1967]. However, in the rest of this chapter and this book we are going to follow von Neumann and Morgenstern and assume, for an n-person game, that

1) players can communicate and form coalitions with other players, and
2) players can make sidepayments to other players.

The resulting theory is known as the theory of *cooperative games with sidepayments*. We will find that the theory of such games is rich and interesting, and applicable in many areas.

In the von Neumann-Morgenstern theory of cooperative games with sidepayments, the questions of major interest are

1) which coalitions should form?
2) how should a coalition which forms divide its winnings among its members?

The information relevant to answering these questions is simply how much each coalition can win if it forms. The specific strategy the coalition will follow to win this amount is not of particular concern. Hence von Neumann and Morgenstern proposed that, just as we abstracted away specific sequences of moves in going from a game in extensive form to a game in normal form, we abstract away specific strategies by going from a game in normal form to a game in *characteristic function form.*

Definition. A *game in characteristic function form* is a set N of players, together with a function v which for any subset $S \subseteq N$ gives a number $v(S)$.

The number $v(S)$, called the *value* of S, is to be interpreted as the amount that the players in S could win if they formed a coalition. The function v is the *characteristic function* of the game. It is traditional to take the value of the empty coalition ϕ (the coalition of no players at all) to be zero.

Any game in normal form can be translated into a game in characteristic function form by taking $v(S)$ to be the *security level* of S. In other words, to calculate $v(S)$, assume that the coalition S forms and then plays optimally under the worst possible condition, which is that all the other players form an opposing coalition $N - S$ and play to hold down the payoff to S. This results in a two-person (S vs. $N - S$) zero-sum game (since $N - S$ is playing to hold down S's payoff). The value of this game to S is $v(S)$. Notice that this is exactly the process we used in our analysis of coalitions in Game 19.1. The characteristic function for Game 19.1, using the symbols R, C, and L for Rose, Colin, and Larry, is

$$v(\phi) = 0$$

$$v(\mathrm{R}) = -4.4 \qquad v(\mathrm{C}) = -4 \qquad v(\mathrm{L}) = -1.43$$

$$v(\mathrm{CL}) = 4.4 \qquad v(\mathrm{RL}) = 4 \qquad v(\mathrm{RC}) = 1.43$$

$$v(\mathrm{RCL}) = 0$$

That the game is zero-sum is reflected in the fact that $v(S) + v(N - S) = 0$ for any subset S of players.

If the original game is not zero-sum, the procedure in the preceding paragraph can still be used to get a game in characteristic function form. However, the resulting characteristic function form game may not be a very accurate reflection of the normal form game, since if a coalition S forms in a non-zero-sum game, it may not be advantageous for $N - S$ to form at all, and if $N - S$ does form, it may not be advantageous for it to play to hold down the payoff to S. We will see an extreme example of this in Chapter 21, where we consider n-person versions of Prisoner's Dilemma.

There is an important relation among the values of different coalitions which holds for all games in characteristic function form which arise from games in normal form:

DEFINITION. A characteristic function form game (N, v) is called *superadditive* if $v(S \cup T) \geqslant v(S) + v(T)$ for any two disjoint coalitions S and T.

If (N, v) arises from a normal form game by taking $v(S)$ to be the security level of S, it will always be superadditive. For if two coalitions S and T, with no common members, decide to join together to form $S \cup T$, they can always assure themselves of at least $v(S) + v(T)$ by simply continuing to do what they would do if they hadn't joined. Of course, they may often be able to do better than this by coordinating their actions.

It is perfectly possible to consider characteristic function form games which do not arise from normal form games. In fact, many interesting situations can be conveniently modeled directly as characteristic function form games. Here are three simple examples.

GAME 19.2: DIVIDE THE DOLLAR [von Neumann and Morgenstern, 1944]. Three players will be given a dollar if they can decide, by majority vote, how to divide the dollar among themselves.

Here we can take $N = \{1, 2, 3\}$ and the characteristic function is

$$v(\phi) = v(1) = v(2) = v(3) = 0$$
$$v(12) = v(13) = v(23) = v(123) = 1$$

GAME 19.3: THE COMMUNICATIONS SATELLITE GAME [McDonald, 1977]. Western Union (W), Hughes Aircraft (H), and General Telephone (G) can put up individual communications satellites, or share satellites jointly in different combinations. McDonald calculates the values of the coalitions (in millions of dollars) as

$$v(\phi) = 0 \qquad v(W) = 3 \qquad v(H) = 2 \qquad v(G) = 1$$
$$v(WH) = 8 \qquad v(WG) = 6.5 \qquad v(HG) = 8.2 \qquad v(WGH) = 11.2$$

This game is non-constant sum, but is superadditive.

GAME 19.4: LEGISLATION IN THE UNITED STATES. To become law, a bill must be approved by a majority of the House of Representatives and a majority of the Senate and must be signed by the President, or it must be approved by $\frac{2}{3}$ of both the House and the Senate (overriding the President's veto).

Here $N = \{$members of the House, members of the Senate, the President$\}$. We can take all coalitions as having one of two values, +1 (winning) or 0 (losing). Then $v(S) = 1$ if and only if S contains at least a majority of both the House and the Senate together with the President, or S contains at least $\frac{2}{3}$ of both the House and the Senate. The game is constant-sum and superadditive.

In the following chapters we will develop techniques for analyzing games 19.1–19.4 and many others.

Exercises for Chapter 19

1. In Game 19.1
   a) Calculate the expected payoffs if all players play their prudential strategies.
   b) If Rose believes that Colin and Larry will both play their prudential strategies, what strategy would be best for Rose? We might call this Rose's *counter-prudential* strategy.
   c) Calculate the counter-prudential strategies for Colin and for Larry.
   d) Is the outcome of all players playing prudentially an equilibrium?
   e) If all players play counter-prudentially instead of prudentially, what will be the result? Who will gain and who will lose?

2. [Brams, 1977a] imagines a situation in which three television networks compete for shares of the viewer market in a prime time slot. ABC has four possible programs it could show, while NBC and CBS each have two. The networks estimate percentage shares of viewers as follows (I've changed Brams' figures a little):

CBS 1

		NBC 1	NBC 2
ABC	1	(35, 34, 31)	(36, 33, 31)
	2	(36, 33, 31)	(37, 31, 32)
	3	(35, 35, 30)	(38, 33, 29)
	4	(34, 34, 32)	(36, 32, 32)

CBS 2

		NBC 1	NBC 2
ABC	1	(38, 33, 29)	(36, 35, 29)
	2	(34, 36, 30)	(38, 34, 28)
	3	(36, 31, 33)	(37, 33, 30)
	4	(35, 34, 31)	(34, 35, 31)

a) Are any of the network's strategies dominated? Does any network have a dominant strategy?
b) Draw a movement diagram. Are there any equilibria?
c) Could you make any predictions about what might happen in this game?

3.

Larry A

		Colin A	Colin B
Rose	A	(4, 3, 3)	(1, 2, 7)
	B	(3, 5, 2)	(0, 4, 6)

Larry B

		Colin A	Colin B
Rose	A	(3, 6, 1)	(2, 5, 3)
	B	(2, 7, 1)	(1, 6, 3)

a) Draw the movement diagram for this constant-sum game. Discuss dominance and equilibria. If this game were played without coalitions, what would you predict would happen?
b) If Colin and Larry form a coalition against Rose, set up and solve the resulting 2×4 game. What will the payoffs to the three players be?
c) Same for Colin vs. Rose-and-Larry.
d) Same for Larry vs. Rose-and-Colin.
e) If no sidepayments are possible, would any player be worse off joining a coalition than playing alone?
f) Draw the diagram showing each player's preferred coalition partner. Conclusions?
g) If sidepayments are possible, write the characteristic function for this game.

4. Find characteristic function form games which model the following situations.
   a) Allen is chair of a committee whose other members are Bates, Curtis and Donahue. The committee operates by majority rule, but if there is a tie, the chair's preference prevails.
   b) Each of Allen, Bates, Curtis and Donahue needs one microcomputer. An individual can buy a microcomputer for $1800, but group discounts are available. Two computers cost $3400, three cost $4800, and four cost $6000. There is an additional $100 discount per computer if all the members of the group are educators. Allen, Bates and Curtis are educators. [Hint: let the value of a coalition be the *savings* that coalition could obtain over buying as individuals. Thus $v(A) = 0$ since Allen by himself can't save over his individual price of $1700. $v(CD) = 100$ since C and D could buy individually at $3500, but jointly at $3400.]

5. Consider the following non-constant-sum game. Rose, Colin and Larry each choose either "A" or "B". The payoff to each player is the total number of B's chosen by *all* players. For example, if the outcome is ABB, the payoffs are (2, 2, 2). Find the characteristic function form of this game, and discuss whether it is a reasonable representation of the game.

20. Application to Politics: Strategic Voting

In the 1980 United States presidential election, there were three candidates: Democrat Jimmy Carter, Republican Ronald Reagan, and Independent John Anderson. In the summer before the election, polls indicated that Anderson was the first choice of 20% of the voters, with about 35% favoring Carter and 45% favoring Reagan. Since Reagan was perceived as much more conservative than Anderson, who in turn was more conservative than Carter, let us make the simplifying assumption that Reagan and Carter voters had Anderson as their second choice, and Anderson voters had Carter as their second choice. We then have the situation

Reagan voters (45%)	Anderson voters (20%)	Carter voters (35%)
R	A	C
A	C	A
C	R	R

If all of these voters voted for their favorite candidate, Reagan would win with 45% of the vote. However, voters need not vote for their favorite candidate. Although it is never advantageous to vote for one's last choice candidate, there are situations in which a vote for a second choice candidate can be helpful. Let us amalgamate voters in each class and think of the situation as a three-person game in which each voter bloc has two strategies:

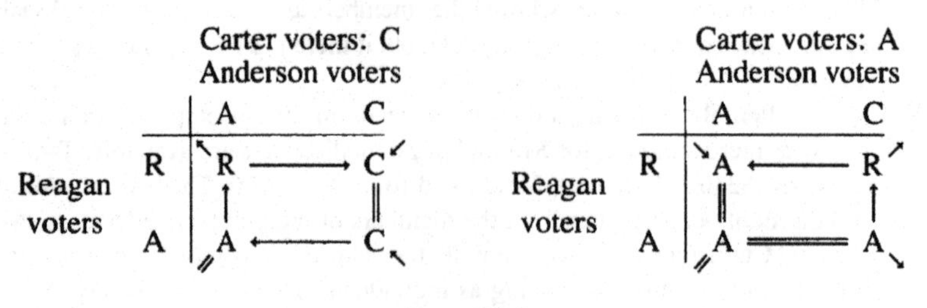

Game 20.1

The movement diagram shows that this game has three equilibria, at RCC (C wins) and at RAA and AAA (A wins). In particular, note that the *sincere* outcome RAC (R wins) is not an equilibrium.

The analysis of this game can be simplified by noticing that the Reagan voters have a dominant strategy of R. Given this, the game reduces to

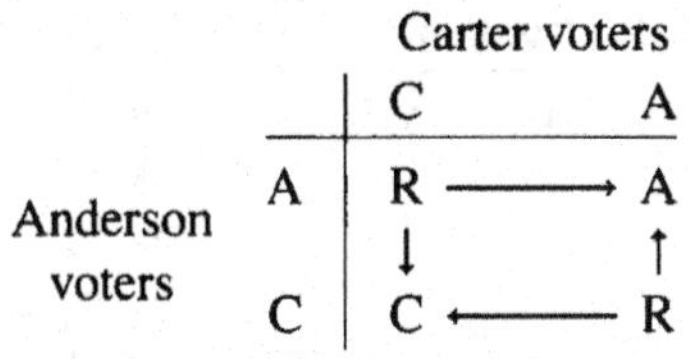

		Carter voters	
		C	A
Anderson voters	A	R ⟶	A
		↓	↑
	C	C ⟵	R

The sincere outcome is at the upper left. Both Carter and Anderson voters could improve upon this outcome by voting for their second choice candidate. In fact, in the summer and fall of 1980 the Carter campaign urged Anderson voters to vote for Carter to keep Reagan from winning. Many liberal Anderson supporters heeded this message and voted for Carter, although Reagan still won the election.

Robin Farquharson [1969] introduced the terminology which has become standard for this kind of situation. A voter's *sincere* strategy is to vote for her first choice candidate. If a voter has a dominant strategy, that strategy is called *straightforward.* Thus Reagan voters have a straightforward strategy, which is to vote sincerely for R. An *admissible* strategy is one which is not dominated. Thus Reagan voters have just one admissible strategy, but Anderson and Carter voters each have two admissible strategies. (Recall that voting for one's last choice candidate is not admissible.) Following an admissible strategy which is not sincere is called *sophisticated voting.* The dilemma of liberal Anderson voters in 1980 was whether to cast sincere votes for Anderson, or sophisticated votes for Carter. When there are three or more candidates in an election, plurality voting often places voters in this kind of game-theoretic dilemma.

Similar situations arise in legislative bodies, which often use versions of *sequential pairwise voting.* In this voting method, two alternatives are voted on first. The majority winner is then paired against the third alternative, the winner of that contest is paired against the fourth alternative, and so on. The order in which alternatives are paired is called the *agenda* of the voting. Let us consider an example.

In March 1988 the United States House of Representatives cast an historic vote which defeated a plan, devised by the House Democratic leadership, to provide humanitarian aid to the U.S.-backed "Contra" rebels in Nicaragua. The vote was unexpected and politically complex. Here is a simplified model of the situation. There were three alternatives.

A: the Reagan administration-supported bill to provide arms to the Contra rebels,

H: the Democratic leadership bill to provide humanitarian aid but not arms,

N: giving no aid to the rebels.

We will reduce the complicated real situation to seven voters with preferences as follows:

CR (2 voters)	MR (1 voter)	MD (2 voters)	LD (2 voters)
A	A	H	N
N	H	A	H
H	N	N	A

The initials stand for Conservative Republicans, Moderate Republicans, Moderate Democrats, and Liberal Democrats. Conservative Republicans, for example, thought that the Democratic bill H was so inadequate that they would rather give no aid than pass H.

In the parliamentary agenda, the first vote was between A and H, with the winner to be paired against N. The result was

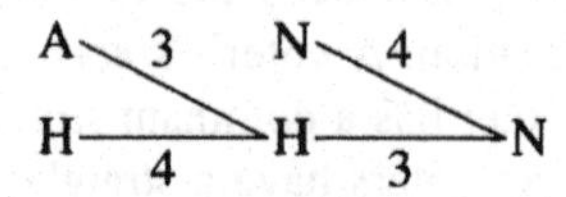

Agenda 1: sincere voting

In the first vote, H got solid Democratic support to beat A (4 to 3 in our model). However, in the following vote on H vs. N, conservative Republicans and liberal Democrats combined to defeat H, resulting in no aid to the Contras.

In our model, the above agenda with sincere voting leads to a victory for N. Would a different outcome be possible? First, let us consider the same agenda, but with sophisticated voting. In the final vote of a pairwise sequence, insincere voting cannot help. In any admissible strategy for the sequence of votes, the final vote will be sincere. Hence if H wins the first vote, we have seen that the final outcome will be N. On the other hand, if A wins the first vote, then in the second vote A will beat N by 5 to 2, so A will be the final outcome. Hence sophisticated voters should view the first vote as a contest between N (represented by H) and A. The Republicans should vote sincerely for A, the liberal Democrats should vote sincerely for H, but the moderate Democrats should have voted sophisticatedly for A. The outcome would have been A instead of N.

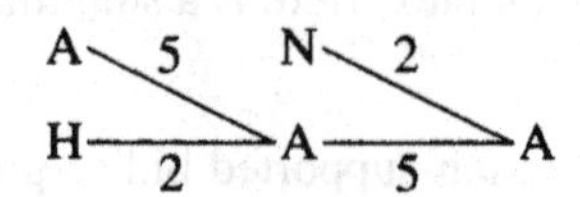

Agenda 1: sophisticated voting

Alternatively, we could consider altering the agenda. In this example an appropriate sequential pairwise agenda could have produced *any one* of the alternatives as the winner under sincere voting:

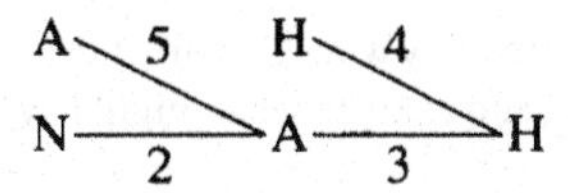

Agenda 2: sincere voting

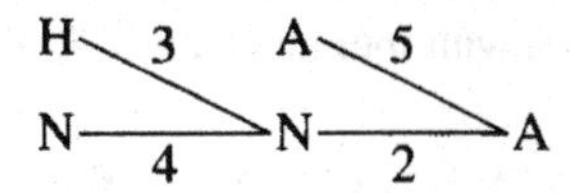

Agenda 3: sincere voting

Exercise 2 asks you to find the sophisticated voting outcome for these agendas. Exercise 3 illustrates an even more drastic case of the effect which the choice of agenda can have on the outcome of sequential pairwise voting.

We have seen that there are situations—they are not at all uncommon—in which both plurality voting and sequential pairwise voting require voters to make sophisticated game-theoretic choices. This is not desirable from the point of view of democratic theory: if voters have incentive to vote insincerely, how can we be sure that the voting outcome will reveal and represent the true preferences of the voters? Surely it would be better to find a voting method (necessarily different from either plurality voting or sequential pairwise voting) which would be *strategyproof*: in all voting situations all voters would have straightforward strategies, which would be their sincere strategies. It would never be advantageous to vote insincerely. However, Gibbard [1973] and Satterthwaite [1975] have proved that when there are three or more alternatives to be voted on, there is only one voting scheme which is strategyproof: to choose one voter and allow him to select the winning alternative. In other words, every voting scheme for three or more alternatives is either *dictatorial* or vulnerable to strategic manipulation. In a democratic society, all voters must be players in n-person games.

Chapter 2 of [Straffin, 1980a] considers a number of other voting methods and illustrates their susceptibility to strategic manipulation. Chapter 3 of that book explores some modern attempts to ameliorate the effects of the Gibbard-Satterthwaite theorem.

Exercises for Chapter 20

1. In the New York Senate election in 1970 there were three candidates: Conservative James Buckley (B), liberal Republican Charles Goodell (G), and Democrat Richard Ottinger (O). Suppose there were three classes of voters with preferences as follows:

Conservatives (39%)	Liberals (28%)	Democrats (33%)
B	G	O
G	O	G
O	B	B

With mostly sincere voting, Buckley won. Construct the appropriate three-person game, draw its movement diagram, and analyze what might have happened with sophisticated voting.

2. Find the sophisticated voting outcomes for Agendas 2 and 3. Which classes of voters would vote insincerely?

3. Consider the voting situation with four alternatives A, B, C, D and three voters with preferences as follows:

Voter 1	Voter 2	Voter 3
A	B	C
B	C	D
C	D	A
D	A	B

Assuming sincere voting, find sequential pairwise agendas which would produce each one of the four alternatives as the final winner. Why might the voters be particularly upset if they happen to use an agenda which produces D as the winner?

4. Set up the game and find optimal sophisticated voting strategies if the Contra aid decision discussed in this chapter had been made by plurality voting instead of by sequential pairwise voting. [You may need to find a way to picture a four-person game in normal form. Alternatively, you could show that players MR and MD have straightforward strategies, so the game reduces to 2×2.]

21. *N*-Person Prisoner's Dilemma

In Chapter 19 I mentioned that not every n-person non-constant-sum game can be analyzed in a satisfactory way using the characteristic function form. In this chapter we will look at a particularly important type of a game which cannot. Game 21.1 is a three person example.

Larry C

		Colin C	Colin D
Rose	C	(1, 1, 1)	(0, 3, 0)
	D	(3, 0, 0)	(2, 2, −2)

Larry D

		Colin C	Colin D
Rose	C	(0, 0, 3)	(−2, 2, 2)
	D	(2, −2, 2)	(−1, −1, −1)

Game 21.1

This game is symmetric for the three players, and strategy D dominates strategy C for all of them. The unique equilibrium is DDD with payoffs $(-1, -1, -1)$. This is a Pareto inferior outcome, since CCC with payoffs $(1, 1, 1)$ would be better for all three players. We recognize the game as a 3-person form of Prisoner's Dilemma.

With the addition of the third player comes the possibility of coalitions, so let us analyze what might happen if Colin and Larry, say, form a coalition. The resulting game is

		Colin-and-Larry CC	CD	DC	DD
Rose	C	(1, 2)	(0, 3)	(0, 3)	(−2, 4)
	D	(3, 0)	(2, 0)	(2, 0)	(−1, −2)

Game 21.2

Notice that Rose's game has a saddle point (by dominating strategies for both players) at DDD, so that Rose's security level is -1. In Colin-and-Larry's game, DCD and DDC are saddle points and the value is zero. By symmetry, the results are the same for other coalitions and we have

$$v(\phi) = 0 \qquad v(\text{R}) = v(\text{C}) = v(\text{L}) = -1$$
$$v(\text{RC}) = v(\text{RL}) = v(\text{CL}) = 0 \qquad v(\text{RCL}) = 3.$$

It looks from this analysis as if it would be better to be in a two-person coalition than to be an individual player (at least if side payments are allowed), and that it would be best of all to join a three-person coalition with everyone agreeing to play C. This result is illusory, an aberration of the characteristic function form. To see this, let us go back and think about what really might happen in Game 21.2.

If the players play prudentially the outcome will be at DCD or DDC, with payoffs $(2, 0)$. Colin-and-Larry would indeed get its security level of 0, but Rose would do much better than her security level of -1. The pessimistic assumption that Colin-and-Larry will play to hold down Rose's payoff is unrealistic here. This means that Rose, or any player, is *better off alone* than in a two-person coalition, or even in the grand coalition of all three players. If you were left in a room with other players to negotiate about how to play this game, the best thing to do would be to run outside and slam the door! The other two players, if they behave rationally, would then conclude that their best strategy would be to play one C and one D and divide their payoffs equally, each getting 0. You, playing D, would win a payoff of 2.

The first moral is that analysis by the characteristic function can be misleading for n-person non-constant-sum games. The second moral is that achieving cooperation is going to be tricky in any situation which can be modeled by an n-person prisoner's dilemma game. To investigate exactly how tricky it can be, consider the following five person game. Each player has two strategies, C and D. The payoff to each player depends on whether that player plays C or D, and also on the number of other players who play C, as follows:

		Number of others choosing C				
		0	1	2	3	4
Player chooses	C	−2	−1	0	1	2
	D	−1	0	1	2	3

payoffs to player

Game 21.3: Altruism

Since each D payoff is larger than the corresponding C payoff, D dominates C for every player. If all players play D, each player gets -1. This is Pareto inferior, since if all players played C, each player would get 2.

Experiment. Find four friends or enemies and play Game 21.3 ten times, keeping track of the choices and payoffs to all players. At the end, calculate the total score for each player. Compare each player's total score to the number of times that player chose C.

If you did this experiment, I think you'll find the results discouraging. To see why, consider the effect of playing C instead of D. First, it will *lower* your score by one unit, regardless of what the other players are doing. (All the C payoffs are one lower than the corresponding D payoffs.) Second, your choice of C will, for everyone else, increase by one the number of other players choosing C, and thereby *raise* their payoff by one. Every choice of C instead of D hurts you and helps everyone else. This is altruism indeed.

When five groups of students in my Game Theory class did this experiment, only 6 of 25 students played C five or more times. Call them *cooperators*. Three of the five groups contained two cooperators each. Call those groups *partially cooperative*. Payoffs worked out as follows:

		Individual	
		Cooperator	Defector
Group	Partially cooperative	−5.7	+4.4
	Uncooperative	—	−3.7
		average payoff to individual	

The way to succeed in this game is to be a defector in a group where some other people are trying to cooperate. In uncooperative groups, everyone fares badly. The way to do worst is to cooperate in groups where defectors take advantage of you. In the world of n-person prisoner's dilemma, virtue is punished and opportunism is rewarded.

An n-person prisoner's dilemma game is characterized as any game in which each of n players has two strategies, call them C and D, such that

i) for every player, D is a dominant strategy, and

ii) if all players choose D, all will be worse off than if all players had chosen C.

In an influential essay, Garrett Hardin [1968] has argued that these characteristics hold for a number of the most important global problems we face at the end of the twentieth century, for example exhaustion of our natural resources, pollution of our environment, and continued population growth. The title of Hardin's essay, "The tragedy of the commons," comes from a description by W. F. Lloyd of a dilemma faced by small farmers in England in the eighteenth century. Picture a community of farmers with access to a common pasture. Each farmer considers the effect of grazing an additional cow in the pasture. The benefit is the profit from an additional cow. The cost is that his additional cow may help to overgraze the commons, but this cost will be spread among all farmers and our farmer's share of it will be small. Since the benefit to the farmer outweighs the cost to him, the farmer rationally chooses to graze another cow, and another and another. So do all the other farmers, and the commons is inexorably destroyed by overgrazing. Hardin quotes Alfred North Whitehead:

> "The essence of dramatic tragedy is not unhappiness. It resides in the solemnity of the remorseless working of things."

In n-person prisoner's dilemma, individual choices of dominant strategies remorselessly yield the Pareto inferior equilibrium. Exercise 3 asks you to work through a numerical example. In Exercise 4, you are invited to consider whether and how population growth, nuclear arms proliferation, and some less crucial problems meet the conditions of n-person prisoner's dilemma. [Schelling, 1978] and Chapter 7 of [Hamburger, 1979] consider a number of other examples.

We close with a pollution game from [Shapley and Shubik, 1969]. Consider five villages around a lake. Each village takes its drinking water from the lake, and discharges its sewage into the lake. Each faces the choice of whether to treat its sewage before discharging it. The cost to a village of treating its sewage is $50,000 per year. On the other hand, the yearly cost for cleaning the water it takes from the lake to make it drinkable, is $20,000 *times* the number of villages which do not treat their sewage. The following table shows the total yearly cost to a village as a function of whether it treats or does not treat its sewage, and of the number of other villages who treat their sewage.

		Number of other villages which treat				
		0	1	2	3	4
Individual village	Treat	13	11	9	7	5
	Don't treat	10	8	6	4	2

cost to the village in $10,000

Since it always costs a village less to pollute the lake than to treat its waste, every village will pollute and each of them will pay $100,000 per year in cleanup costs. This is considerably more than the $50,000 it would cost each village to keep the lake unpolluted.

Is there any remedy? We know that no individual village will profit by treating its waste, and in fact its taxpayers would pay an extra $30,000 per year to save money for people in the *other* villages. However, perhaps several villages could join in a common water district. Two villages cannot save themselves money like this, but three villages can, since if all three treated their waste each would save $60,000 in cleanup costs, while only paying $50,000 in waste treatment costs. However, we are being naive if we expect such a coalition to form easily. The reason, as we have seen before, is that the two villages not included in the coalition would reap the $60,000 benefit without paying the $50,000 cost. All of the villages would like the other villages to join in a common water district, but would try hard not to join themselves. They would all prefer to be "free riders."

Hardin suggests that the remedy in this kind of situation is for all the villages jointly to agree to *force* all the villages to join in a common coalition. After all, such an application of common force would benefit every village. He argues that "mutual coercion mutually agreed upon" must be the central principle of a new morality, if we are to deal with the n-person prisoner's dilemmas which threaten the future of our planet. If you think about how such a morality might apply to the problem of population growth, you can see how controversial this suggestion is. If you wish to avoid it, you must argue that the problems Hardin considers are not serious, or that they are not prisoner's dilemmas, or you must suggest an alternative way to deal with the dilemma. In any case, the model of n-person prisoner's dilemma has a central place in any serious thought about contemporary global issues.

Exercises for Chapter 21

1. In ten plays of Game 21.3, suppose that c is the number of C's played by all five players ($0 \leqslant c \leqslant 50$), and c_i is the number of C's played by player i ($0 \leqslant c_i \leqslant 10$). Show that player i's total payoff is $c - 2c_i - 10$. (Hence player i would like to have c large, but c_i small.)

2. Consider the following cooperation-defection game, which is not a five-person prisoner's dilemma since it is not always advantageous to play D.

		Number of others choosing C				
		0	1	2	3	4
Player chooses	C	−3	−1	1	3	5
	D	−1	0	1	2	3

payoffs to player

   a) Suppose in Round 1 all players play C. Would a C player think it beneficial to switch to D in the next round?
   b) Suppose in Round 1 the players play four C's and one D (CCCCD). Would a C player want to switch to D in the next round? Would the D player want to switch to C?
   c) Repeat the analysis for initial plays of CCCDD, CCDDD, CDDDD, and DDDDD.
   d) What would you expect to happen in this game?
   e) Can you think of any real life situations like this?

3. Six farmers live around a common which can support six cows of value \$1000 each. For each additional cow beyond six which is grazed on the common, the value of every cow grazed on the common decreases by \$100. Each farmer has two strategies

   C: graze one cow on the commons

   D: graze two cows on the commons.

   a) Fill in the table

		Number of others choosing C					
		0	1	2	3	4	5
Farmer chooses	C						
	D						

value of Farmer's cows

   and show that this game fulfills the conditions of 6-person prisoner's dilemma.

   b) What is the smallest size coalition of farmers which could benefit by having all its members choose C? (Assume that the farmers not in the coalition would continue to pursue their individual self interest by choosing D.)
   c) If such a coalition formed, would you prefer to be in it or not in it?

4. Discuss which of the following situations could reasonably be modeled as n-person prisoner's dilemmas. What are the strategies C and D? Is an individual player better off playing D regardless of what other players do? Are all players better off if they all play C than if they all play D?
   a) Nations developing and stockpiling nuclear weapons.
   b) Couples in a poor country deciding whether to have another child.
   c) Deciding whether or not to turn on your air conditioner on a peak summer day.
   d) Standing on your seat at a rock concert.

22. Application to Athletics: Prisoner's Dilemma and the Football Draft

In an n-person Prisoner's Dilemma game, if all players rationally pursue their own best interests, all players end up worse off than if they had all followed some individually less rational line of play. We have seen that this type of situation appears often in the course of human interactions. One surprising place it can appear is in sequential choice procedures. In this chapter we will analyze one well-known sequential choice procedure, the professional football draft in the United States.

In football, basketball and other professional sports, teams choose new players by a draft system which involves sequential choices. From the pool of available players, the team with the worst win-loss record in the previous season gets first choice. The second worst team gets second choice, and so on until all teams have chosen. The procedure is then repeated for as many rounds as needed until all teams have exhausted their choices. Presumably this system, by giving the worst teams priority in the selection of new players, makes the teams more competitive next season, thereby fostering public interest in the sport.

The problem with this seemingly admirable procedure is that it is subject to a sophisticated form of Prisoner's Dilemma. To begin to see why, let us first consider the choice process in its simplest context. Suppose there are just two teams, the Blues and the Reds, and there are four players in the draft, whom we will unimaginatively call A, B, C, and D. Suppose the teams' preferences for the players are as follows:

Blues	Reds
A	B
B	C
C	D
D	A

Thus the Blues most want player A, and least want player D, while the Reds most want B and least want A. Such preferences, by the way, are not at all unreasonable. Player A might be a fine quarterback, but the Reds, if they already have a superb quarterback, could be quite uninterested. If both teams follow a *sincere* drafting strategy, and the Blues go first, the result will be

	Blues	Reds
1st round	(A) —	(B)
	B	C
2nd round	(C) —	(D)
	D	A

Sincere choices

Each team will end up with its first and third choices. However, just as voters need not vote sincerely, teams need not choose sincerely, and they may be able to benefit from insincere choices. Suppose in this situation that

i) the teams know each other's preference orders, and

ii) each team's goal is to get the players it most prefers (not, for example, to keep the other team from getting the players that team wants).

In the sports draft context, assumption i) is quite realistic, since teams have extensive knowledge both of the abilities of players in the draft and of other teams' needs. Assumption ii) may or may not be valid in real contexts—teams have been known to draft players they don't need in order to keep other teams from getting them!

Given these assumptions, the Blues can benefit by noticing that the Reds would never willingly draft player A. Hence the Blues could safely draft their second choice player B in Round One, and plan on picking up player A in Round Two. The result is

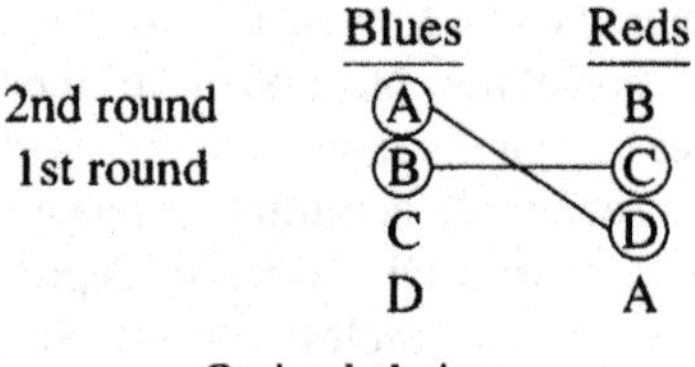

Optimal choices

The Reds have no defense against this strategy consistent with assumption ii). These choices are optimal for both teams.

In this example there is no Prisoner's Dilemma. The Reds have lost but the Blues have gained, and the resulting outcome is Pareto optimal. In fact, it is possible to design an algorithm for optimal choices in a two-team draft (see Exercise 1) and prove that the resulting outcome is always Pareto optimal [Brams and Straffin, 1979]. When only two teams are drafting, a Prisoner's Dilemma cannot arise.

However, as soon as there are three teams drafting, the situation changes. Consider the following example with three teams and six players:

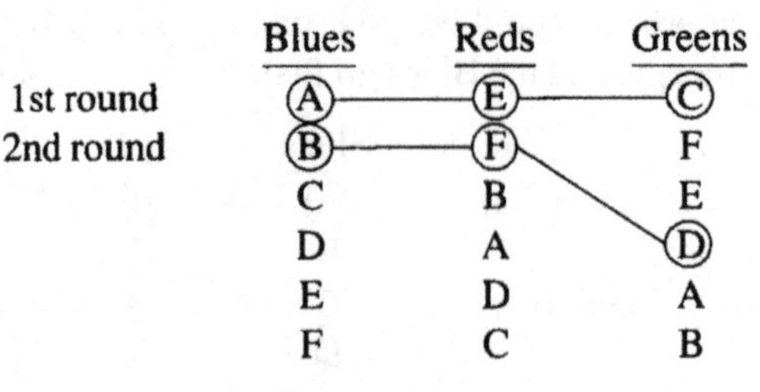

Sincere choices

With the sincere choices shown, the Blues and the Reds both get their top two players, and even the Greens don't do too badly. The Blues and the Reds obviously have nothing to gain from insincerity, but let us consider the Greens. If the Blues and the Reds choose sincerely in Round One, the Greens could choose F instead of C, and the result would be

Blues Reds Greens
1st round
2nd round
A E C
B F F
C B E
D A D
E D A
F C B

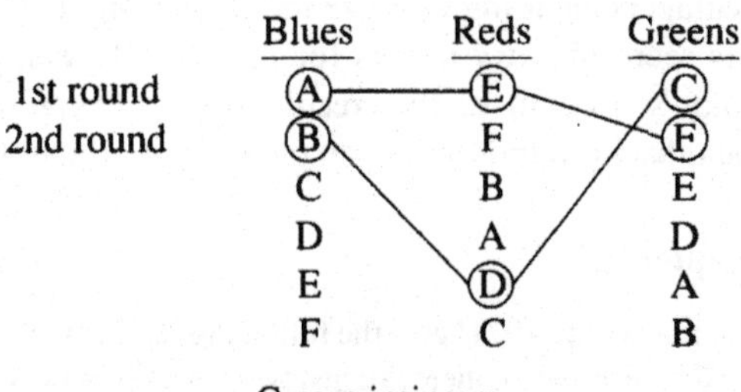

Greens insincere

The Greens certainly benefit by this insincerity. Now consider the Reds, who will be hurt by Green insincerity if they choose E in Round One. Do they have any defense? It turns out that choosing F in Round One will not help (Exercise 2), but choosing B will:

Blues Reds Greens
1st round
2nd round
A E C
B F F
C B E
D A D
E D A
F C B

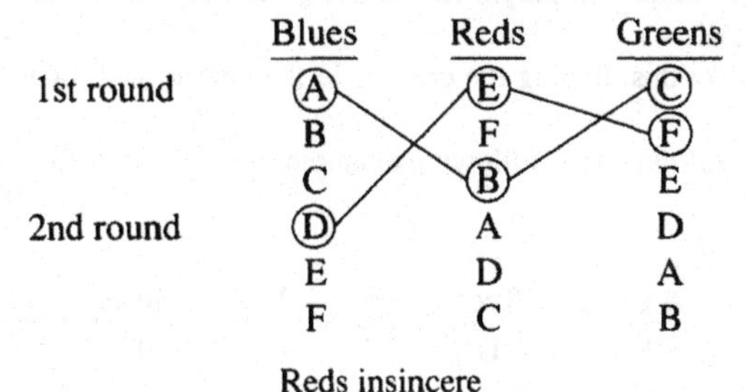

Reds insincere

The reason is that the Greens must now choose C in Round One, or lose him to the Blues. Now the Reds' defensive insincerity has hurt the Blues. Do the Blues have a defense? Choosing B in Round One will not help (Exercise 3), but choosing C will produce

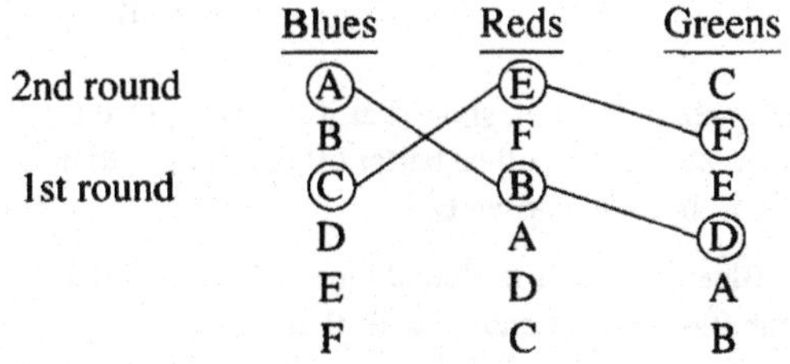

Blues Reds Greens
2nd round
1st round
A E C
B F F
C B E
D A D
E D A
F C B

Optimal choices

It turns out—see [Brams and Straffin, 1979] for a proof—that these choices are optimal for all of the teams.

If we compare the result of these optimal choices with the result of the original sincere choices, we see the Prisoner's Dilemma. All of the teams are strictly worse off than if they had all chosen sincerely. Rational analysis in the pursuit of self-interest has hurt everyone. The participants in the sophisticated choice scheme of the player draft are just as vulnerable to the paradox of Prisoner's Dilemma as the villagers deciding to graze their cows on the Commons.

If capital letters seem like dry names for football players, you might enjoy reading Dr. Crypton's account of this result in [Dr. Crypton, 1986]. He has provided names and backgrounds and even pictures of all the players.

Exercises for Chapter 22

1. [Kohler and Chandrasekaran, 1971] give the following "from the bottom up" algorithm for finding optimal choices when there are just two teams (say the Blues and the Reds) drafting:
   i) Under optimal play, the Reds' choice in the *last* round will be the player who is last on the Blues' preference list. Mark that player as the Reds' last round choice and (in your imagination) cross him off both teams' lists.
   ii) The Blues' choice in the last round will be the player who is last on the Reds' reduced list. Mark that player as the Blues' last round choice, and cross him off both teams' lists.
   iii) Continue like this, finding the choices in the next-to-last round, and on up to the first round.

   Try out this algorithm on the following examples, and compare the results with sincere choices.

a)

Blues	Reds
B	D
C	B
E	F
F	C
D	A
A	E

b)

Blues	Reds
F	D
H	F
G	G
C	C
E	B
A	E
D	H
B	A

2. In the 3-team example in the text, show that if in Round One the Blues choose A and the Reds choose F, the Greens will be better off choosing E than C. The Reds will end up worse off than if they had chosen E.

3. Show that if the Blues choose B in Round One, the optimal Red response would be to choose A, and the Blues would end up with B and D.

4. Suppose that, with the same preferences, the Reds chose first in the draft, followed by the Greens and then the Blues. Find the sincere choices. It is possible to prove that in this ordering, the sincere choices are in fact optimal. Is it true that a team necessarily benefits by having the first choice in the draft?

5. a) Construct an example with 6 teams drafting 12 players which has the same property as the 3-team example, that the result of optimal choices is worse for all teams than the result of sincere choices. [Hint: can you use the 3-team example?]
   b) Construct a 4-team, 8-player example with the same property. [Hint: start from the 3-team example. This is a real challenge.]

23. Imputations, Domination, and Stable Sets

In this chapter we begin the search for solutions to n-person games in characteristic function form. I should tell you at the outset that the situation is going to be at least as murky as it is for two-person non-zero-sum games. For n-person games there are a number of different useful and illuminating ideas of what a solution might be, but none of them is completely satisfactory in all situations.

Suppose we have an n-person game in characteristic function form (N, v). We will assume that the game is superadditive. The two questions we would like to answer about such a game are

i) which coalition or coalitions should form?

ii) how should a coalition which forms divide its winnings among its members?

Von Neumann and Morgenstern [1944] first observed that the second question might be the most important. For one thing, how different potential coalitions propose to divide their payoffs will strongly influence which coalitions individual players will prefer to join, and hence which coalitions will form. For another, an individual player's ultimate interest is the payoff he gets from the game: coalition membership is a means to that end. Hence von Neumann and Morgenstern proposed concentrating first on the distribution of payoffs to the n players in the game. This will be an n-tuple of numbers $(x_1, x_2, \ldots, x_n)$, where x_i is the payoff to the ith player. There are two important properties this n-tuple should have:

Individual rationality. $x_i \geqslant v(i)$ for all i.

Since $v(i)$ is the amount player i can assure himself without the cooperation of anyone else, it would not be rational for player i to accept less than this as his payoff.

Collective rationality. $\sum_{i \in N} x_i = v(N)$.

Since our game is superadditive, $v(N)$ is the maximal amount which can be divided, so we cannot have $\sum_{i \in N} x_i > v(N)$. If $\sum_{i \in N} x_i < v(N)$, it would be possible for the players to increase all of their payoffs by forming the grand coalition N. Collective rationality says that they should do this. (In Chapter 29 we will look at a solution concept which does not assume collective rationality.)

We have seen these ideas for two-person games in Chapter 11. Collective rationality is just the Pareto Principle. In the Divide-the-Dollar game 19.2, individual rationality requires that $x_i \geqslant 0$ for $i = 1, 2, 3$, and collective rationality requires that $x_1 + x_2 + x_3 = 1$.

DEFINITION. A payoff n-tuple satisfying individual rationality and collective rationality is called an *imputation* for the game (N, v).

To "solve" an n-person game, we should say which imputation or set of imputations we think should arise as the result of rational plays of the game. There is (only) one case in which the answer is easy.

Definition. A superadditive n-person game (N, v) is *inessential* if

$$\sum_{i \in N} v(i) = v(N).$$

Otherwise the game is *essential*.

If (N, v) is inessential, then superadditivity implies that $v(S) = \sum_{i \in S} v(i)$ for all coalitions S. In other words, any coalition can only assure for its members what they could already get on their own. Hence there is no incentive for any coalition to form, and we would expect each player to end up with exactly $v(i)$. In our new terminology, the *only* imputation for an inessential game is the one with $x_i = v(i)$ for all i.

For any essential game there are infinitely many imputations, and for further analysis it is very helpful to have a convenient way of picturing all of them. For 3-person games, such a picture arises from a theorem of elementary plane geometry.

Theorem. Let P be any point in an equilateral triangle of height h. Then the sum of the distances from P to the three sides of the triangle is equal to h.

Proof. Exercise 1a.

To represent the set of all imputations for Divide-the-Dollar, for example, draw an equilateral triangle of height 1, and label its vertices by the names of the players 1, 2, 3. For any point in the triangle, let x_i be the distance to the side *opposite* vertex i, as in Figure 23.1a. Then each point in or on the boundary of the triangle corresponds to a unique triple of numbers (x_1, x_2, x_3) with $x_i \geqslant 0$ and $x_1 + x_2 + x_3 = 1$, i.e., to an imputation of the game. Figure 23.1b illustrates

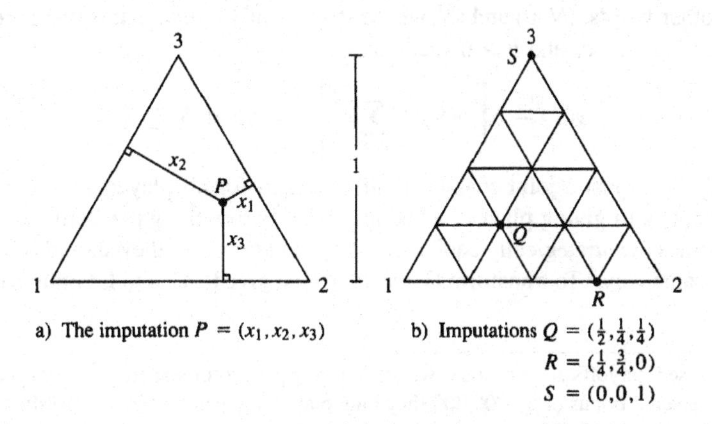

a) The imputation $P = (x_1, x_2, x_3)$

b) Imputations $Q = (\frac{1}{2}, \frac{1}{4}, \frac{1}{4})$
$R = (\frac{1}{4}, \frac{3}{4}, 0)$
$S = (0, 0, 1)$

Figure 23.1: Imputations for Divide-the-Dollar.

three imputations. Notice that player i prefers imputations which are close to vertex i.

What about other three-person games? First, notice that the set of imputations for a game depends only on $v(1), \ldots, v(n)$ and $v(N)$. Hence the triangle of height 1 represents the set of imputations for any three person game with $v(1) = v(2) = v(3) = 0$ and $v(123) = 1$. For example, it represents the imputations for

$$v(1) = v(2) = v(3) = 0$$
$$v(12) = 1/4 \qquad v(13) = 1/2 \qquad v(23) = 3/4$$
$$v(123) = 1,$$

Game 23.1

which is superadditive and essential but non-constant-sum.

Now suppose we have any superadditive, essential three person game, for example Game 19.1:

$$v(\mathrm{R}) = -4.4 \qquad v(\mathrm{C}) = -4 \qquad v(\mathrm{L}) = -1.43$$
$$v(\mathrm{RC}) = 1.43 \qquad v(\mathrm{RL}) = 4 \qquad v(\mathrm{CL}) = 4.4$$
$$v(\mathrm{RCL}) = 0.$$

I claim that we can transform any such game into a game with $v(\mathrm{R}) = v(\mathrm{C}) = v(\mathrm{L}) = 0$ and $v(\mathrm{RCL}) = 1$ by changes in utility scales which do not affect the strategic nature of the game.

DEFINITION. Two games (N, v) and (N, w) are *strategically equivalent* if we can change v into w by

i) for some or all players i, adding a constant c_i to the payoff of player i (and hence to the value of any coalition containing player i), and
ii) multiplying all payoffs by a positive constant.

In other words, (N, v) and (N, w) are strategically equivalent if there are constants $c_1, c_2, \ldots, c_n$ and $b > 0$ such that

$$w(S) = b\left[v(S) + \sum_{i \in S} c_i\right] \qquad \text{for all } S \subseteq N.$$

Change ii) is just a joint change in utility scales for all players, and change i) corresponds to giving player i a "bonus" of c_i before the game starts. Hence if two games are strategically equivalent, one can argue that they should be played in the same way.[†] To transform Game 19.1, first give R, C, and L bonuses of 4.4,

[†]Of course if payoffs are monetary, we are ignoring psychological effects. If players have just received a bonus of $1,000,000 they may play a $10 game more recklessly. Or they might play a $1,000,000 game differently from a $10 game.

4, and 1.43, respectively. Then we would have

$$\begin{aligned} v(\text{R}) &= v(\text{C}) = v(\text{L}) = 0 \\ v(\text{RC}) &= 1.43 + 4.4 + 4 = 9.83 \\ v(\text{RL}) &= 4 + 4.4 + 1.43 = 9.83 \\ v(\text{CL}) &= 4.4 + 4 + 1.43 = 9.83 \\ v(\text{RCL}) &= 4.4 + 4 + 1.43 = 9.83. \end{aligned}$$

Now divide all payoffs by 9.83, and we find that Game 19.1 is strategically equivalent to Divide-the-Dollar! If you think about doing this process in general, the following results should be clear:

- Any superadditive essential n-person game G is strategically equivalent to a unique game with $v(1) = v(2) = \cdots = v(n) = 0$, $v(N) = 1$, and $0 \leqslant v(S) \leqslant 1$ for all $S \subseteq N$. This unique game is called the *0–1 normalization* of G.
- The 0–1 normalization of any essential three person *constant-sum* game is Divide-the-Dollar.

Thus our triangle will picture the imputations for the 0–1 normalization of any three person essential game. Exercise 1b implies that the imputations for the 0–1 normalization of any four-person essential game can be pictured as the points inside or on the boundary of a regular tetrahedron. (For a game with $n > 4$ players, we would need the higher dimensional analogue of a triangle and tetrahedron called an *n-dimensional simplex.*) Often it will be convenient not to do the scale change to make $v(N) = 1$, but just to picture the imputations of a three-person game in a triangle of height $h = v(N)$.

Now that we can picture the imputations in at least a three person game, we still need to say which imputation or imputations we should expect from rational play. To do that, of course, we need to look more closely at the play of a game in characteristic function form. The simplest example is Divide-the-Dollar. I have played and watched my students play many coalition games, and the following abstract captures some of what happens in these games. The players are our friends Rose, Colin, and Larry.

Rose: The fairest way to divide the dollar is surely $(\frac{1}{3}, \frac{1}{3}, \frac{1}{3})$. Since it isn't evenly divisible in cents, why don't we play odd-finger to see who gets the extra penny?

Colin: That may be fair, but there's nothing in the rules which says we have to be fair. I suggest that Larry and I, since we form a majority, agree to split the dollar 50–50.

Larry: Fine with me.

Rose: I'm sorry you don't value fairness, but I can play at this game too. Larry and I would also form a majority, and I'm not as greedy as Colin.

I would be happy to offer Larry 60 cents and keep only 40 cents for myself.

Colin: I don't need to be left out in the cold. I'll offer Larry 70 cents to come with me.

Rose: I'll offer him 80.

Colin: I'll offer him 90.

Larry: Fine with me!

Rose: Wait a minute, Colin. At this point you would be getting only 10 cents and Larry hasn't done anything except accept fat offers. Why don't you and I get together and split the dollar 50–50?

Colin: Good idea.

Larry: Wait a minute! I'll offer 60 cents to Rose!

....

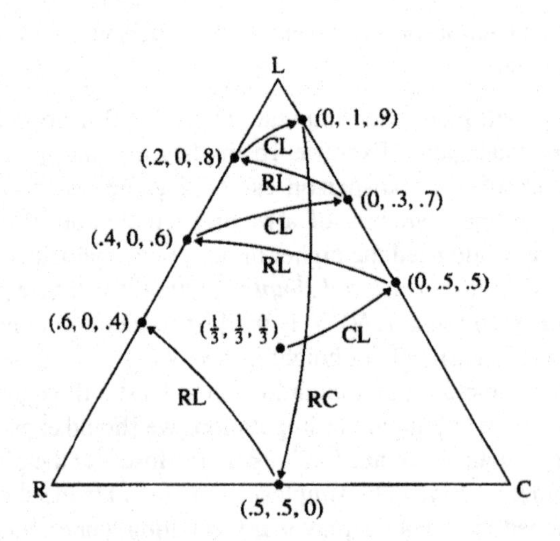

Figure 23.2: Bargaining in Divide-the-Dollar.

Figure 23.2 shows how the proposed imputations move about the triangle. Each transition between imputations is labeled by the name of the coalition which can effect that transition.

Notice that two things are necessary for a transition to occur. First, all members of the coalition which makes the transition must prefer the new imputation to the old, or they would not have incentive to change. For example, when {CL} makes the transition from (40, 0, 60) to (0, 30, 70) both C and L gain. Second, the coalition which makes the transition must be able to enforce the new imputation, which in Divide-the-Dollar means it must be a majority coalition. Von Neumann and Morgenstern abstracted these conditions to make the following

Definition. An imputation $\mathbf{x} = (x_1, \ldots, x_n)$ *dominates* an imputation $\mathbf{y} = (y_1, \ldots, y_n)$ if there is some coalition S such that

i) $x_i > y_i$ for all i in S, and

ii) $\sum_{i \in S} x_i \leqslant v(S)$.

We say that the domination is *via S*, and write $\mathbf{x}\ \mathrm{dom}_S\ \mathbf{y}$. In the scenario of Figure 23.2 each new imputation dominates the preceding one, via the coalition which labels the arrow.

If $\mathbf{x}$ dominates $\mathbf{y}$ via a coalition S, we could expect that S would try to enforce $\mathbf{x}$ over $\mathbf{y}$ and hence that $\mathbf{y}$ would not be a viable outcome for the game. However, the relation of domination is subtle, since there are many different coalitions which can effect it. Three rather strange properties of domination are already present in Divide-the-Dollar:

1) Given two imputations $\mathbf{x}$ and $\mathbf{y}$, it is possible that neither one dominates the other. For example, take $\mathbf{x} = (.50, .50, 0)$ and $\mathbf{y} = (.50, 0, .50)$. Then $\mathbf{x}$ does not dominate $\mathbf{y}$, since only C prefers $\mathbf{x}$ to $\mathbf{y}$, and $v(C) = 0 < .50$. Similarly, $\mathbf{y}$ does not dominate $\mathbf{x}$.
2) Cycles of domination are possible. We may have imputations $\mathbf{x}$, $\mathbf{y}$, and $\mathbf{z}$ such that $\mathbf{y}\ \mathrm{dom}\ \mathbf{x}$, $\mathbf{z}\ \mathrm{dom}\ \mathbf{y}$, and $\mathbf{x}\ \mathrm{dom}\ \mathbf{z}$. For example, if $\mathbf{x} = (.50, .50, 0)$, $\mathbf{y} = (0, .60, .40)$, and $\mathbf{z} = (.30, 0, .70)$, then $\mathbf{y}\ \mathrm{dom}_{\{CL\}}\ \mathbf{x}$, $\mathbf{z}\ \mathrm{dom}_{\{RL\}}\ \mathbf{y}$, and $\mathbf{x}\ \mathrm{dom}_{\{RC\}}\ \mathbf{z}$. In mathematical terms, the relation of domination fails, rather badly, to be transitive. Of course, if there is a cycle of domination, different coalitions must effect the dominations in the cycle.
3) It is possible that *every* imputation is dominated by some other imputation. In fact, this is true in Divide-the-Dollar, as is clear from the diagrams in Figure 23.3.

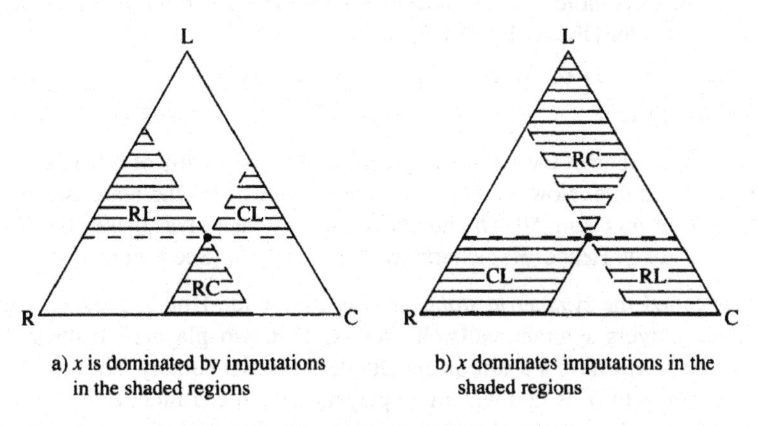

a) x is dominated by imputations in the shaded regions

b) x dominates imputations in the shaded regions

Figure 23.3: Domination in Divide-the-Dollar.

Faced with the fact that, at least in Divide-the-Dollar, every imputation is dominated by some other imputation, and hence any proposed imputation will be unstable, what shall we do? Von Neumann and Morgenstern proposed that we consider not individual imputations, but *sets* of imputations which exhibit two kinds of stability.

DEFINITION. A *stable set* (sometimes called a *von Neumann-Morgenstern solution*) for a game G is a set $\mathcal{J}$ of imputations such that

i) $\mathcal{J}$ is *internally stable*: no imputation in $\mathcal{J}$ is dominated by any other imputation in $\mathcal{J}$, and

ii) $\mathcal{J}$ is *externally stable*: every imputation not in $\mathcal{J}$ is dominated by some imputation in $\mathcal{J}$.

The idea is that if we start at some imputation not in $\mathcal{J}$, external stability will lead us into $\mathcal{J}$. Once in $\mathcal{J}$, internal stability will keep us from moving within $\mathcal{J}$. Hence we might predict that the outcome of the game will be some imputation in $\mathcal{J}$, although we cannot say which specific imputation it will be.

However, there is a problem. It may well be true (and always is true in Divide-the-Dollar) that an imputation in $\mathcal{J}$ will be dominated by an imputation outside of $\mathcal{J}$. When we get in $\mathcal{J}$, there would seem to be no reason to stay there. Von Neumann and Morgenstern handle this objection in an ingenious way. They suggest that we might think of the set $\mathcal{J}$ of imputations as constituting a *social norm* for resolving the distribution problem represented by the game. Thus imputations in $\mathcal{J}$ could be assumed to be inherently more attractive than imputations not in $\mathcal{J}$. Hence a coalition S thinking about replacing an imputation $\mathbf{x}$ in $\mathcal{J}$ by an imputation $\mathbf{y}$ not in $\mathcal{J}$ would be reluctant to do so even if $\mathbf{y}$ dominates $\mathbf{x}$ via S. For there is guaranteed to be (by external stability) an imputation $\mathbf{z}$ in $\mathcal{J}$ which dominates $\mathbf{y}$, and S could foresee the likelihood that $\mathbf{y}$ would be replaced by $\mathbf{z}$, and $\mathbf{z}$ does *not* dominate $\mathbf{x}$ (by internal stability).

To understand more fully the idea of a stable set, it is illuminating to look at the stable sets for Divide-the-Dollar.

THEOREM. $\mathcal{F} = \{(.50, .50, 0),\ (.50, 0, .50),\ (0, .50, .50)\}$ is a stable set for Divide-the-Dollar.

PROOF. First, none of these three imputations dominates any of the others, so $\mathcal{F}$ is internally stable. Now for any imputation $\mathbf{y}$ not in $\mathcal{F}$, there are two players who each get less than .50. The imputation in $\mathcal{F}$ which gives these players each .50 dominates $\mathbf{y}$. Hence $\mathcal{F}$ is externally stable as well. (See Figure 23.4a.)

$\mathcal{F}$ is called the *symmetric stable set* for Divide-the-Dollar, because it treats the three players symmetrically. It predicts that two players—it doesn't say which two—will form a coalition and divide the dollar equally. The social norm corresponding to $\mathcal{F}$ is that when two players form a coalition, a 50–50 split is more natural than an unequal split. Since this corresponds well with our intuition,

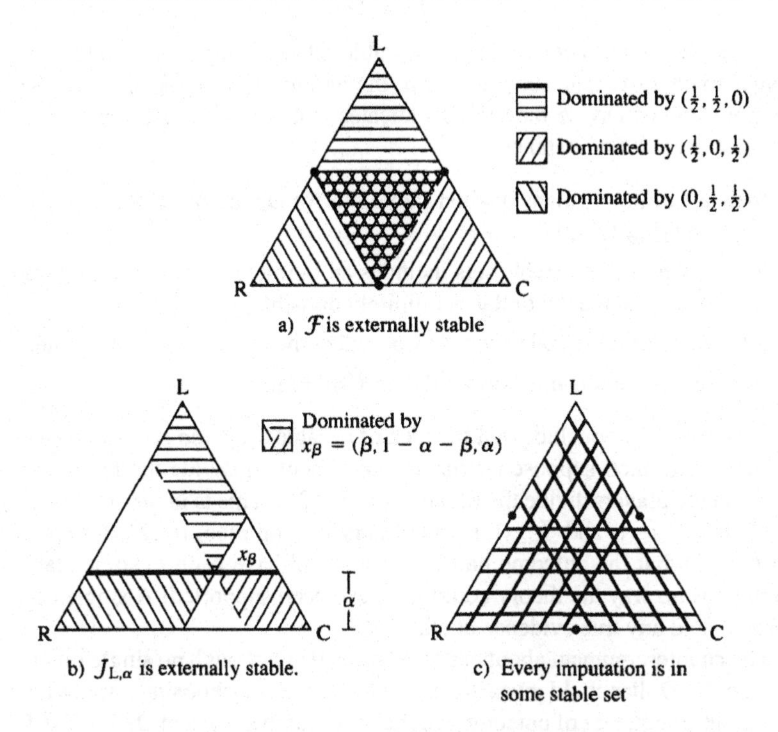

Figure 23.4: Stable sets for Divide-the-Dollar.

$\mathcal{F}$ seems attractive as a "solution" to Divide-the-Dollar. Interestingly, though, there are other stable sets for Divide-the-Dollar.

THEOREM. Let α be any number such that $0 \leqslant \alpha < \frac{1}{2}$. Let $\mathcal{J} = \mathcal{J}_{L,\alpha} = \{$all imputations in which L gets $\alpha\}$. Then $\mathcal{J}$ is a stable set for Divide-the-Dollar.

PROOF. Since L is indifferent between any two imputations in $\mathcal{J}$, and R and C have opposite preferences among the imputations in $\mathcal{J}$, the set is internally stable. Figure 23.4b shows that $\mathcal{J}$ is externally stable. As the imputation $\mathbf{x}_\beta$ moves along the line $\mathcal{J}_{L,\alpha}$, the area it dominates sweeps out the entire imputation triangle, with the exception of the line $\mathcal{J}_{L,\alpha}$ itself.

$\mathcal{J}_{L,\alpha}$ is called a *discriminatory stable set*. The social norm corresponding to it prescribes that Larry has an *a priori* claim on a fraction α of the dollar. Rose and Colin grant this claim, and bargain about how to divide the rest, and there is no social expectation that they will divide it equally. You might think of Larry as the Church in medieval Europe, collecting its tithe of 10% and leaving other sectors of society to fight over the remaining 90%. Notice that the "discrimination" of $\mathcal{J}_{L,\alpha}$ toward L could be favorable: α could be larger than $\frac{1}{3}$. On the other hand, we could have $\alpha = 0$, with Larry being completely "tabu."

If we want to consider the idea of a stable set as giving a solution for n-person games in characteristic function form, the nature of the stable sets for the very simple game of Divide-the-Dollar indicates that we can expect this solution to be complicated. For instance,

- there are infinitely many distinct stable sets for Divide-the-Dollar: $\mathcal{F}$ and $\mathcal{J}_{R,\alpha}$, $\mathcal{J}_{C,\alpha}$, and $\mathcal{J}_{L,\alpha}$ for all $0 \leq \alpha < \frac{1}{2}$.
- all but one of these stable sets includes an infinite number of imputations (at least if we assume the dollar is infinitely divisible).
- all but one of these stable sets do not reflect the symmetry of the game.
- *every* imputation is in some stable set. (See Figure 23.4c.)

One might easily conclude that the idea of a stable set is too general to be useful at all. For instance, suppose we run an experiment on Divide-the-Dollar, and we find that the players divide the dollar (.40, .35, .25). This is in the stable set $\mathcal{J}_{R,.40}$ (and also in $\mathcal{J}_{C,.35}$ and $\mathcal{J}_{L,.25}$), so the theory is supported, right? But recall that no matter what the outcome had been, it would have been in some stable set, so there is no way the theory could *not* have been supported! A single outcome cannot yield any true evidence at all.

The counterargument about experiments is that although no single outcome to Divide-the-Dollar could give evidence that players are choosing imputations in a stable set, a sequence of outcomes could. If successive plays of Divide-the-Dollar yielded outcomes of (.40, .35, .25), (.60, .15, .25), (0, .75, .25), and (.20, .55, .25), we would have support for $\mathcal{J}_{L,.25}$. More generally, von Neumann and Morgenstern argued that the multiplicity of stable sets is a cogent argument *for* the acceptance of this idea as a solution for n-person games. Reality is complicated, and societies have many different customs and norms. The fact that the simple logical structure of the stable set idea leads to a variety of stable sets in practice just means that we may have found an idea powerful enough to reflect reality.

The theory of stable sets was central to mathematical research in game theory in the 1950's and 1960's. Evidence gradually accumulated that the variety and nature of stable sets could be far more complicated and pathological than von Neumann and Morgenstern had suspected. Lloyd Shapley, for instance, discovered that it is possible to take quite a general bounded set in the plane—your signature, for example—and construct a five-person game for which that set appears as one piece of a stable set. There are games for which all stable sets are so "infinitely flaky" that they cannot be computed by any finite procedure. There are games for which no stable set reflects the symmetries of the game. It became more and more difficult to justify these kinds of complexities by von Neumann and Morgenstern's argument. Finally, William Lucas [1968] constructed a ten-person game for which there is *no* stable set. (See [Owen, 1982].) By then the situation had become so messy that Martin Shubik [1982] describes Lucas' result as coming "almost as a relief to those working in this field." It effectively ended any lingering belief that stable set theory should be taken as the general solution theory for n-person characteristic function games. If you would like to read more about developments in stable set theory, [Lucas, 1981] and [Shubik, 1982] are good places to start.

In subsequent chapters we will investigate a series of other solution theories for n-person games: the core (Chapter 25), the Shapley value (Chapter 26), Aumann-Maschler bargaining sets (Chapter 29), the nucleolus (Chapter 31). None of these theories is uniformly satisfactory as a general solution theory for all games, but all of them offer interesting insights and all of them answer some questions about some games very well. Von Neumann and Morgenstern stable sets should be considered in this context. In some cases they are pathologically complicated, in some cases they do not exist, but in some cases they offer valuable insights. I think that Divide-the-Dollar, with its symmetric stable set and its variety of discriminatory stable sets, is one such case.

Exercises for Chapter 23

1. a) Prove that if P is any point in an equilateral triangle of height h, then the sum of the perpendicular distances from P to the sides of the triangle is equal to h. [Hint: Join P to the vertices of the triangle and consider areas.]
   b) Prove that if P is any point in a regular tetrahedron of height h, then the sum of the perpendicular distances from P to the four faces of the tetrahedron is equal to h. (Hence imputations for a four-person essential game can be represented as points in a regular tetrahedron.)

2. a) Find the 0–1 normalization of the game

$$v(\phi) = 0 \qquad v(1) = 1 \qquad v(2) = 2 \qquad v(3) = 3$$
$$v(12) = 5 \qquad v(13) = 7 \qquad v(23) = 9 \qquad v(123) = 12$$

   b) What happens if you try to 0–1 normalize the following game? Why?

$$v(\phi) = 0 \qquad v(1) = 1 \qquad v(2) = 2 \qquad v(3) = 3$$
$$v(12) = 3 \qquad v(13) = 4 \qquad v(23) = 5 \qquad v(123) = 6$$

3. Consider the following imputations for Game 23.1.

$$\mathbf{p} = (\tfrac{1}{4}, \tfrac{3}{8}, \tfrac{3}{8}) \qquad \mathbf{r} = (\tfrac{1}{4}, \tfrac{3}{4}, 0) \qquad \mathbf{t} = (\tfrac{1}{2}, \tfrac{1}{4}, \tfrac{1}{4})$$
$$\mathbf{q} = (\tfrac{5}{16}, \tfrac{9}{16}, \tfrac{1}{8}) \qquad \mathbf{s} = (\tfrac{1}{2}, \tfrac{1}{2}, 0)$$

   a) Show that $\mathbf{p}$ dominates $\mathbf{t}$.
   b) Show that $\mathbf{q}$ dominates $\mathbf{r}$.
   c) Find an imputation in the list which dominates $\mathbf{s}$.
   d) Show that neither $\mathbf{p}$ nor $\mathbf{q}$ is dominated by any imputation in the list.
   e) Find an imputation not in the list which dominates $\mathbf{q}$.
   f) Argue that no imputation, in the list or not, can dominate $\mathbf{p}$.

4. Consider the four-person constant-sum game

$$v(\phi) = v(A) = v(B) = v(C) = v(D) = 0$$
$$v(AB) = 50 \qquad v(CD) = 70 \qquad v(AC) = 30$$
$$v(BD) = 90 \qquad v(AD) = 30 \qquad v(BC) = 90$$
$$v(ABC) = v(ABD) = v(ACD) = v(BCD) = v(ABCD) = 120$$

and imputations

$$\mathbf{u} = (25, 25, 35, 35) \qquad \mathbf{y} = (35, 65, 10, 10)$$
$$\mathbf{w} = (20, 40, 20, 40) \qquad \mathbf{x} = (30, 60, 30, 0).$$

Find a cycle of domination among these imputations.

5. Since Game 19.1 is strategically equivalent to Divide-the-Dollar, we can translate stable sets for Divide-the-Dollar into stable sets for Game 19.1.
   a) Do this for the stable set $\mathcal{F}$. Interpret the result as specifying how each two person coalition should divide its winnings, if it is the two person coalition which forms.
   b) Do this for the discriminatory stable set $\mathcal{J}_{L,1/4}$. If this stable set is the social norm, how much should Rose and Colin allot to Larry?

6. For the "apex" game

$$v(\phi) = v(A) = v(B) = v(C) = v(D) = 0$$
$$v(BC) = v(BD) = v(CD) = 0$$
$$v(AB) = v(AC) = v(AD) = 1$$
$$v(ABC) = v(ABD) = v(ACD) = v(BCD) = v(ABCD) = 1$$

Show that the following four imputations form a stable set:

$$\mathbf{p} = (\tfrac{2}{3}, \tfrac{1}{3}, 0, 0) \qquad \mathbf{r} = (\tfrac{2}{3}, 0, 0, \tfrac{1}{3})$$
$$\mathbf{q} = (\tfrac{2}{3}, 0, \tfrac{1}{3}, 0) \qquad \mathbf{s} = (0, \tfrac{1}{3}, \tfrac{1}{3}, \tfrac{1}{3}).$$

7. Show that for a superadditive inessential game

$$v(S) = \sum_{i \in S} v(i) \qquad \text{for all } S \subseteq N.$$

24. Application to Anthropology: Pathan Organization

One of the earliest applications of cooperative game theory in anthropology was Fredrik Barth's study of political incentives among the Yusufzai Pathans of northern Pakistan. Barth used a simple game-theoretic model to clarify the logical structure of these incentives and explain behaviors which at first seemed puzzling. I'll describe the situation as it was when he studied it in the 1950's.

The Yusufzai Pathans (pronounced "Puh-táns") occupied the agriculturally rich lowlands in and around the Lower Swat Valley, near the Khyber Pass in the Northwest Frontier Province of Pakistan. Almost all land was owned by a dominant male aristocracy, and their land tenure system was complicated. An individual Pathan inherited a specified fraction of his village estate, but not particular fields. Rather, he was allotted fields of dry and irrigated land corresponding to the size of his share, and to ensure fairness, allotments were changed every five or ten years. This re-allotment, and any land dispute, was decided by a village council. Since land was the basis of economic and social status among the Pathans, and control of land was temporary and under the control of a local political body, this system ensured that politics was central to Pathan society.

To protect and enhance his individual status, then, a Pathan needed political allies. The inheritance system ensured that those allies would *not* come from his closest relatives, since relatives occupied nearby fields, and hence the most intense land disputes tended to be between close relatives. The search for political allies which must not be close relatives produced a division of Pathan society into two blocs or factions which dominated political life:

> Small descent segments of fathers and sons, or brothers, align under recognized leaders in a two-faction split which extends throughout the Yusufzai and bordering areas. The blocs function in protecting the interests of their members by exerting underhand pressure, by working together in the councils, and as armies in the case of fighting. The relative strength of the opposed blocs, and the importance of the contestants to their respective allies, are more important in settling conflicts than are abstract principles of justice. [Barth, 1959]

The blocs were important in social life as well:

> Every village is divided in two wards, and political opponents occupy different wards. The wards, not the whole villages, are the operative political and economic units. There is a headman or chief in every ward. This office is coveted by many pretenders. It offers great personal and strategic advantages to the incumbent. [Barth, 1959]

Since individual Pathans were free to change their bloc allegiance, bloc membership could change over time, and the dynamics of change were the focus of Barth's study:

> Balance between the blocs is maintained, according to the Pathan conception of the system, by the essential cupidity of politicians. The stronger bloc in an area will tend to grow in numbers and land by making good their advantage over the weaker bloc. Inevitably, however, rivalry will develop between the leaders of the growing alliance, until one such leader sees his chance to capture supreme control of the territory by seceding with his followers and joining the weaker bloc, which thereby becomes the stronger. [Barth, 1959]

With this background, we can ask a number of interesting questions, for example

1) Since bloc membership is based not on loyalty to the bloc, but on self-interest, and it is clearly advantageous to be in the majority bloc, why doesn't the majority bloc grow more quickly than it does?
2) Since there seem to be no moral scruples about exploiting members of the weaker bloc, why is the weaker bloc not quickly bankrupted?
3) Since there is no stigma attached to defecting from a bloc (in fact, strategic defectors seem to be admired), why is it that defections are relatively rare, and the bloc structure is relatively stable?

In other words, there was a stability in Pathan society which appeared puzzling because it was not enforced by any moral sanctions. Barth believed that this stability could be explained entirely in terms of incentives inherent in the social structure. To illustrate how these incentives work, he proposed a simple five person game model. Here is an adaptation of Barth's model.

There are five players A, B, C, D, and E, who have different amounts of disposable wealth a, b, c, d, and e, which can be taken from them and given to other players by majority vote. In the natural course of events, players A, B, and C form a coalition against players D and E, appropriate their wealth $(d + e)$, and distribute it equally among themselves. The resulting payoffs for the five players would be

$$\left(\frac{d+e}{3}, \frac{d+e}{3}, \frac{d+e}{3}, -d, -e\right). \tag{24.1}$$

Now suppose D decided to join the majority bloc. That would leave only E to be exploited, and his wealth would have to be split four ways, giving payoffs

$$\left(\frac{e}{4}, \frac{e}{4}, \frac{e}{4}, \frac{e}{4}, -e\right). \tag{24.2}$$

D would benefit, but A, B, and C would all lose. In the language of game theory, (24.1) dominates (24.2) via coalition ABC in the five person majority game. The reason that the majority bloc does not grow quickly is evident: its original members would be very reluctant to accept defectors from the minority. This kind of reasoning has been generalized and applied widely in political science by William Riker, who calls it the *size principle*: "Assuming perfect information in zero-sum games, the equilibrium size of a winning coalition is always minimal." See [Riker and Ordeshook, 1973] for an extended discussion.

To address our second and third questions, we need to add an additional feature to our model. Recall that each bloc has a leader, and that this leadership post is coveted, not so much for economic reasons as for social status. In the model, we will suppose that each of the two blocs has a leader, who gets an additional payoff l for each of his followers. If A is the leader of bloc ABC and D is the leader of DE, then the total payoffs are

$$\left(\frac{d+e}{3}+2l,\ \frac{d+e}{3},\ \frac{d+e}{3},\ -d+l,\ -e\right). \tag{24.3}$$

Now suppose that D and E tempt C's defection by offering him the leadership of the new majority bloc CDE. The new payoffs would be

$$\left(-a+l,\ -b,\ \frac{a+b}{3}+2l,\ \frac{a+b}{3},\ \frac{a+b}{3}\right). \tag{24.4}$$

We are interested in when (24.4) would dominate (24.3). E would certainly prefer (24.4). D would prefer (24.4), and hence would be willing to offer C leadership, only if

$$\frac{a+b}{3} > -d+l, \qquad \text{i.e.} \qquad l < \frac{a+b}{3}+d. \tag{24.5}$$

Finally, C would prefer the new payoffs, and hence would accept the offer, only if

$$\frac{a+b}{3}+2l > \frac{d+e}{3}, \qquad \text{i.e.} \qquad l > \frac{(d+e)-(a+b)}{6}. \tag{24.6}$$

Evidence from Pathan society indicates that l is fairly large, so that unless the members of the minority bloc are exceptionally rich, condition (24.6) is probably often satisfied. The offer of leadership would be a strong temptation to a member of the majority bloc. Hence the important constraint is probably (24.5): would the leader of the minority bloc be willing to give up his leadership?

The leader A of the majority bloc has a strong incentive to insure that inequality (24.5) is not satisfied. What control does he have? He can't control l, and he probably can't control $(a+b)$, since in this context $(a+b)$ would refer to D's estimate of how much wealth could be milked from A and B. However, he does have control over d, which in this context would be the actual amount of wealth his coalition is milking from D. A would want to keep this amount from becoming too large. If he is successful, he will prevent offers leading to defections from his bloc.

We have answers to questions (2) and (3). The majority does not exploit the minority too heavily because doing so would result in offers to potential defectors in the majority bloc. In the absence of heavy exploitation, offers to defectors are not made because the leader of the minority bloc is not willing to sacrifice his leadership position. Barth cites evidence that "the interest of the chiefs in limiting the opposition between blocs is proverbial," and the model shows clearly why this should be so. It also gives us the additional insight that the social status accorded to bloc leaders is functional in Pathan society, since it acts to ensure relative political stability.

Notice that in this application we have not used any sophisticated ideas about solutions to an N-person game. Sometimes merely formulating a model of a strategic situation as a game can be an important step in understanding that situation.

Exercises for Chapter 24

1. Redo the analysis of the conditions for defection if there are seven players A through G, A is the leader of ABCD, E is the leader of EFG, and E tempts D to defect.
2. Suppose we have $2n + 1$ players, organized in blocs of size $n + 1$ and n. For simplicity suppose that all wealths are equal to 1. The leadership bonus is l per follower. What are the necessary conditions for defection?

25. The Core

Von Neumann and Morgenstern's idea of a stable set (Chapter 23) was historically the first proposed solution for games in characteristic function form. It was introduced because in essential constant-sum games, no single imputation is stable: every imputation is dominated by some other imputation. However, in non-constant-sum games there may be undominated imputations, and in the early 1950's Gillies and Shapley pointed out that the set of all undominated imputations in a game is an object worthy of study. Gillies called it the *core* of the game.

Definition. The *core* of a game in characteristic function form is the set of all undominated imputations, i.e., the set of all imputations $(x_1, \ldots, x_n)$ such that for all $S \subseteq N$, $\sum_{i \in S} x_i \geqslant v(S)$.

Recall that the definition of an imputation specifies that $x_i \geqslant v(i)$ for all players i (individual rationality), and that $\sum_{i \in N} x_i \geqslant v(N)$ (collective rationality). Hence we can think of the core condition as extending these ideas to *coalitional rationality* for coalitions of size between 1 and n. An imputation is in the core if the members of any coalition S are getting, in total, at least as much as S could guarantee them.

Of course, the core may be the empty set of no imputations at all, and indeed it is the empty set for any essential constant-sum game. For an example of a three-person game with non-empty core, consider Game 23.1:

$$v(1) = v(2) = v(3) = 0$$

$$v(12) = \tfrac{1}{4} \qquad v(13) = \tfrac{1}{2} \qquad v(23) = \tfrac{3}{4}$$

$$v(123) = 1$$

An imputation (x_1, x_2, x_3) is in the core of this game if

$$\begin{aligned} x_1 + x_2 &\geqslant v(12) = \tfrac{1}{4} \\ x_1 + x_3 &\geqslant v(13) = \tfrac{1}{2} \\ x_2 + x_3 &\geqslant v(23) = \tfrac{3}{4}. \end{aligned}$$

For example, $(\frac{1}{4}, \frac{1}{2}, \frac{1}{4})$ meets these conditions, so is in the core. To find all imputations in the core, we can use the imputation diagram, as in Figure 25.1. Notice that

$$\begin{aligned} x_1 + x_2 \geqslant \tfrac{1}{4} &\text{ if and only if } x_3 \leqslant \tfrac{3}{4} \\ x_1 + x_3 \geqslant \tfrac{1}{2} &\text{ if and only if } x_2 \leqslant \tfrac{1}{2} \\ x_2 + x_3 \geqslant \tfrac{3}{4} &\text{ if and only if } x_1 \leqslant \tfrac{1}{4}. \end{aligned}$$

When we graph these inequalities, they have a non-empty intersection in the imputation triangle. The core is a trapezoid with vertices at $(0, \frac{1}{4}, \frac{3}{4})$, $(0, \frac{1}{2}, \frac{1}{2})$,

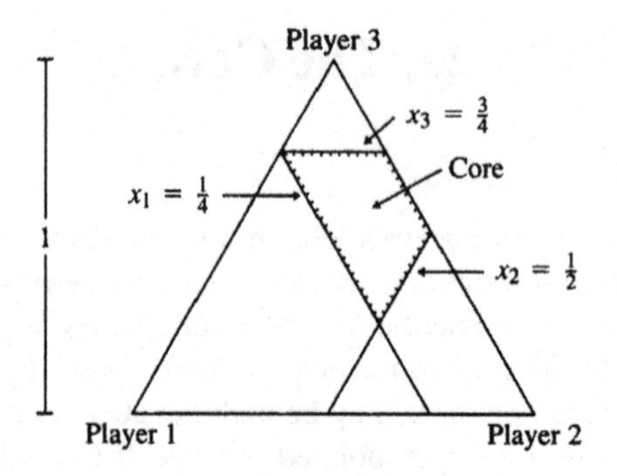

Figure 25.1: The core of Game 23.1.

$(\frac{1}{4}, 0, \frac{3}{4})$, and $(\frac{1}{4}, \frac{1}{2}, \frac{1}{4})$. Any imputation in this trapezoid is not dominated by any other imputation in the triangle.

To illustrate properties and uses of the core, we will work out two other examples. First, consider a game analyzed by von Neumann and Morgenstern. Player 1 has a house which she values at \$100,000, and offers it for sale. There are two potential buyers, players 2 and 3, who each have \$200,000 in cash, and value the house at \$200,000. Here is the resulting game, with values measured in units of \$100,000:

$$v(1) = 1 \qquad v(2) = 2 \qquad v(3) = 2$$
$$v(12) = 4 \qquad v(13) = 4 \qquad v(23) = 4$$
$$v(123) = 6$$

The value of coalition $\{12\}$, for example, is derived by noting that if players 1 and 2 form a coalition, player 1 can sell the house to player 2 for some price p, $1 \leqslant p \leqslant 2$. Player 1 will then have p units in cash, and player 2 will have $2 - p$ units plus a house he values at 2 units. Hence the total wealth of the coalition is 4 units. If coalition $\{23\}$ forms, no sale is possible. If coalition $\{123\}$ forms, player 1 sells the house to either player 2 or player 3.

When we normalize this game we get

$$v(1) = v(2) = v(3) = 0$$
$$v(12) = v(13) = 1 \qquad v(23) = 0$$
$$v(123) = 1$$

Game 25.1

The core of this game is the set of imputations (x_1, x_2, x_3) with $x_1 + x_2 \geqslant 1$, $x_1 + x_3 \geqslant 1$, and $x_2 + x_3 \geqslant 0$. The only imputation which satisfies these conditions is $(1, 0, 0)$, so in this game the core is a single point (see Figure 25.2b). This unique undominated imputation corresponds to player 1 selling her house to one of the two buyers at the maximum possible price of \$200,000. Player 1 gets *all* the benefit of economic cooperation.

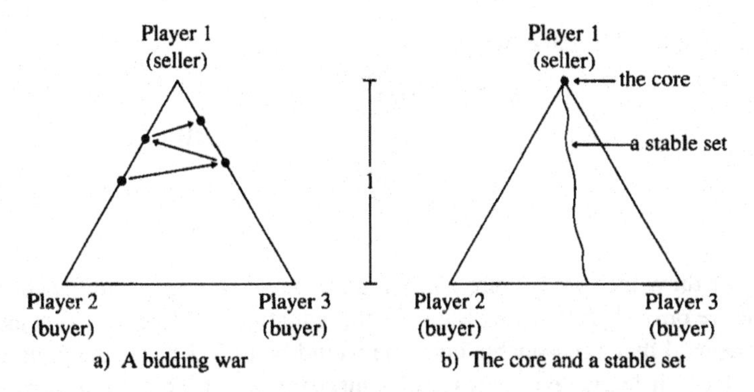

Figure 25.2: Von Neumann and Morgenstern's real estate game 25.1.

We might imagine the sequence of events leading to this outcome as a bidding war between the two buyers:

Player 2: "I offer \$150,000."

Player 3: "I offer \$160,000."

Player 2: "I offer \$170,000."

Player 3: "I offer \$180,000..."

See Figure 25.2a. As long as the price is less that \$200,000, both buyers have incentive to continue raising each other's bid. If you have an object of interest to just one buyer, you have to haggle. If you have an object of interest to two buyers, you can play them off against each other.

Von Neumann and Morgenstern point out that $(1, 0, 0)$ by itself is not a von Neumann-Morgenstern stable set. In fact, $(1, 0, 0)$ doesn't dominate *any* imputation. One stable set for this game is shown in Figure 25.2b. It consists of the core point together with a curve which runs from the core to the bottom of the imputation triangle (any curve which always slants downward at a 60° angle or steeper will do—see Exercise 3). This kind of stable set has a nice economic interpretation. We imagine players 2 and 3 getting tired of trying to outbid each other, and engaging in a bit of collusion. They will negotiate jointly with player 1 and buy the house at some price $1 \leqslant p \leqslant 2$. Player 2, say, will then get the house and pay the price p, but he will give player 3 a sidepayment, the amount of which is agreed on in advance for all prices p, for his cooperation in avoiding a bidding war. In this example, the core captures the effect of competition between the buyers, and the stable sets show the possible effects of collusion.

Next, consider McDonald's communications satellite game 19.3:

$$v(G) = 1 \qquad v(H) = 2 \qquad v(W) = 3$$

$$v(GH) = 8.2 \qquad v(GW) = 6.5 \qquad v(HW) = 8$$

$$v(GHW) = 11.2$$

If we normalize this game, we get

$$v(G) = v(H) = v(W) = 0$$
$$v(GH) = 5.2 \qquad v(GW) = 2.5 \qquad v(HW) = 3$$
$$v(GHW) = 5.2$$

Game 25.2

When we draw the core (Figure 25.3), it turns out to be empty, but the core conditions are only slightly inconsistent. In this situation, William Lucas suggested to McDonald that a reasonable outcome would be (2.3, 2.8, .1), the point in the center of the little triangle where the core just barely isn't. This would correspond to GTE and Hughes sharing a satellite, but giving a small sidepayment to Western Union to keep it from trying to disrupt their coalition. We will return in Chapter 31 to this idea of how to proceed when the core is empty.

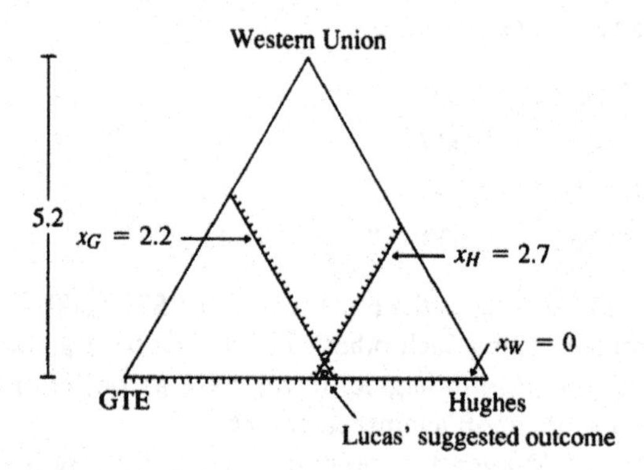

Figure 25.3: The communications satellite game 25.2 has an empty core.

Exercises for Chapter 25

Some stationery stores sell graph paper ruled in equilateral triangles instead of squares. Such triangular graph paper is very helpful for these exercises and some in later chapters.

1. For the game $v(1) = v(2) = v(3) = 0$, $v(12) = 5$, $v(13) = 2$, $v(23) = 3$, $v(123) = 6$, draw the core in the imputation triangle. What is its shape? List the imputations at its vertices.

2. For the microcomputer buying game of Exercise 19.4b, which of the following imputations are in the core? If an imputation is not in the core, give one coalition S whose members are getting less than $v(S)$.
   a) (225, 225, 225, 225) b) (600, 100, 100, 100)
   c) (300, 300, 300, 0) d) (500, 200, 200, 0)
   e) (100, 500, 300, 0) f) (200, 400, 0, 300)
   g) (250, 250, 250, 150) h) (150, 150, 150, 450)
   Do you think that all the imputations in the core are "fair"?

3. Check that the stable set for Game 25.1 shown in Figure 25.2b really is a stable set, i.e. that it is both internally stable and externally stable. The easiest argument is geometric. Start by choosing an arbitrary imputation in the set and shading all the imputations which it dominates. (Careful!)

4. Change the real estate example by supposing that players 2 and 3 each have $300,000 in cash, and the house is worth $200,000 to player 2, but $300,000 to player 3.
   a) Write the characteristic function form for this game, using units of $100,000. Then normalize it to make $v(1) = v(2) = v(3) = 0$.
   b) Draw the core in the imputation triangle. Interpret the core in terms of what would happen in the sales situation.

5. The Glove Market. Three traders L_1, L_2, and L_3 each have one lefthand glove, and two traders R_1 and R_2 each have one righthand glove. The value of a coalition is the number of *pairs* of gloves it has. E.g. $v(LRR) = 1$, $v(LL) = 0$, $v(LLLRR) = 2$.
   a) Show that the imputation $(\frac{2}{5}, \frac{2}{5}, \frac{2}{5}, \frac{2}{5}, \frac{2}{5})$ (payoffs to L_1, L_2, L_3, R_1, and R_2, respectively) is dominated by some other imputation.
   b) Show that the imputation $(\frac{2}{10}, \frac{2}{10}, \frac{2}{10}, \frac{7}{10}, \frac{7}{10})$ is dominated by some other imputation.
   c) The core of this game consists of a single imputation. What is it? Verify that for this imputation, every coalition gets at least its value.

6. Lake Pollution. Recall the example of Chapter 21. There are five villages around a lake. Each village may pay 5 units to treat its sewage, or put the sewage into the lake untreated. For each village which pollutes the lake, each village must pay 2 units to treat its intake water. In Chapter 21 we considered briefly the possible effects of coalitions of villages. Now we will look at this in more detail.
   a) If a coalition of k villages forms, compute the payoff to the coalition if its members all clean up their wastes, and if they all pollute. Assume that villages not in the coalition will pollute. Complete the following table:

Size of coalition	If clean up	If pollute
1	−13	−10
2		−20
3	−27	
4		
5		

   [The payoff for a 3-village coalition which cleans up is computed as follows. Each village pays 5 units to clean its waste, plus 4 units to clean its intake water, since the two villages not in the coalition are still polluting. Hence the total cost to the coalition is $3 \times 9 = 27$.] What size coalitions will find it less costly to clean up than to pollute?
   b) Let the value of a coalition of size k be the payoff obtained by following its least costly strategy from a). This defines a game. Normalize the game so that each one-person coalition has value 0. [Hint: the grand coalition will have value 25.]
   c) Show that the following imputations are in the core of the game: $(5, 5, 5, 5, 5)$, $(13, 3, 3, 3, 3)$, $(11, 11, 1, 1, 1)$. Try to interpret each of these in terms of the original situation. Who is cleaning up, and who is paying for it?

7. Prove that any essential constant sum game with $n > 1$ players has an empty core. [Hint: reduce the game to zero-one normal form, and consider the $(n-1)$-player coalitions.]

8. The zero-one normal form of a general superadditive 3-person game is

$$v(1) = v(2) = v(3) = 0$$
$$v(12) = a \qquad v(13) = b \qquad v(23) = c$$
$$v(123) = 1$$

with $0 \leqslant a, b, c \leqslant 1$. Find a general condition on a, b, c for the core of the game to be non-empty.

26. The Shapley Value

The two solution concepts for n-person games in characteristic function form which we have considered so far, von Neumann-Morgenstern stable sets and the core, are based on possible divisions of $v(N)$ which might result from coalitional bargaining. In particular, they are based on the concept of domination. The core is the set of undominated imputations. A stable set is a more complicated set of imputations meeting more subtle conditions of internal and external stability. In both cases the solution is a set of imputations rather than a single imputation, since coalitional bargaining is too complex for us to expect to be able to predict a single outcome.

Instead of asking about the possible results of actual coalitional behavior, suppose we ask if there might be a single imputation which would, in some sense, represent a *fair* distribution of payoffs. Such an imputation might never arise from the competitive behavior of coalitions, but it would be the imputation which an outside arbitrator might impose, taking into account the relative strengths of the various coalitions. For example, in Divide-the-Dollar we have seen that a likely outcome might be that one of the three two-person coalitions would form and divide the dollar equally between its two members, resulting in one of the three imputations $(\frac{1}{2}, \frac{1}{2}, 0)$, $(\frac{1}{2}, 0, \frac{1}{2})$, or $(0, \frac{1}{2}, \frac{1}{2})$ in the stable set $\mathcal{F}$. Given the symmetry of this game, the fair division is certainly $(\frac{1}{3}, \frac{1}{3}, \frac{1}{3})$, although we might not expect this outcome to arise in any play of the game.

In 1953 Lloyd Shapley gave a general answer to this fair division question, which has come to be known as the *Shapley value* of a game in characteristic function form. Shapley first wrote down three axioms which he argued capture the idea of a fair distribution of payoffs. The situation is that for a game with n players and characteristic function v, we are going to assign an imputation $\varphi = (\varphi_1, \ldots, \varphi_n)$.

AXIOM 1. φ should depend only on v, and should respect any symmetries in v. That is, if players i and j have symmetric roles in v, then $\varphi_i = \varphi_j$.

AXIOM 2. If $v(S) = v(S - i)$ for all coalitions $S \subseteq N$, that is, if player i is a *dummy* who adds no value to any coalition, then $\varphi_i = 0$. Furthermore, adding a dummy player to a game does not change the value of φ_j for other players j in the game.

The third axiom is based on the idea of the sum of two games. Suppose that (N, v) and (N, w) are two games with the same player set N. Then we can define the *sum game* $v + w$ by simply defining $(v + w)(S) = v(S) + w(S)$ for all coalitions S. Since we now have three different games under consideration, denote by $\varphi[v]$, $\varphi[w]$, and $\varphi[v + w]$ the imputations we will assign to those games. Then

AXIOM 3. $\varphi[v + w] = \varphi[v] + \varphi[w]$.

The idea is that if it is fair for some player i to get $\varphi_i[v]$ in v and $\varphi_i[w]$ in w, it would seem fair for him to get the sum of these two payoffs in the game $v + w$.

Shapley was then able to prove the following beautiful result.

THEOREM [Shapley, 1953]. There is one and only one method of assigning an imputation φ to a game (N, v) which satisfies Axioms 1, 2, and 3.

Sketch of the proof: Shapley showed that the three axioms force a unique imputation φ by cleverly breaking down an arbitrary game into a sum of symmetric games with dummies added. We will illustrate his method on the game

$$v(A) = v(B) = v(C) = 0$$
$$v(AB) = 2 \qquad v(AC) = 4 \qquad v(BC) = 6$$
$$v(ABC) = 7,$$

Game 26.1

and it will be clear that the method works for any game. First, a bit of notation. For any coalition $S \subseteq N$, let v^S be the game defined by

$$v^S(T) = \begin{cases} 1 & \text{if } T \supseteq S \\ 0 & \text{if } T \not\supseteq S \end{cases}$$

In other words, in v^S, a coalition can win 1 unit if and only if it contains all of the members of S. Notice that all the players in S have symmetric roles in v^S, and all the players not in S are dummies. Hence by Axioms 1 and 2 we must have

$$\varphi_i[v^S] = \begin{cases} 1/s & \text{if } i \in S \\ 0 & \text{if } i \notin S \end{cases}$$

where s is the number of players in S. Now, starting with the original game v, consider the new game

$$v' = v - 2v^{\{AB\}} - 4v^{\{AC\}} - 6v^{\{BC\}}.$$

For this game, we have

$$v'(A) = v'(B) = v'(C) = v'(AB) = v'(AC) = v'(BC) = 0$$
$$v'(ABC) = 7 - 2 - 4 - 6 = -5.$$

In other words, $v' = -5v^{\{ABC\}}$, which yields

$$v = 2v^{\{AB\}} + 4v^{\{AC\}} + 6v^{\{BC\}} - 5v^{\{ABC\}}.$$

Hence by Axiom 3, we must have

$$\varphi_A[v] = 2 \times \tfrac{1}{2} + 4 \times \tfrac{1}{2} + 6 \times 0 - 5 \times \tfrac{1}{3} = 1\tfrac{1}{3}$$
$$\varphi_B[v] = 2 \times \tfrac{1}{2} + 4 \times 0 + 6 \times \tfrac{1}{2} - 5 \times \tfrac{1}{3} = 2\tfrac{1}{3}$$
$$\varphi_C[v] = 2 \times 0 + 4 \times \tfrac{1}{2} + 6 \times \tfrac{1}{2} - 5 \times \tfrac{1}{3} = 3\tfrac{1}{3}.$$

This unique imputation has been forced by the axioms. The construction works in general, and Shapley's theorem is proved.

The construction in the proof of the theorem gets complicated as games get larger, and the Shapley value would be hard to compute in that way. However, a combinatorial analysis of the construction gives a different, and intuitively very appealing, way of computing φ, which was also given in [Shapley, 1953]:

> Consider the players forming the grand coalition step by step, starting with one player, then adding a second player and continuing until N is formed. As each player joins, award to that player the value he adds to the growing coalition. The resulting awards give an imputation. However, there are $n!$ different orders in which the grand coalition might form. Average the imputations given by all of these orders. This averaged imputation is φ.

To clarify this process, let's see how it works for Game 26.1. The calculation is as follows:

	Value added by		
Order	A	B	C
ABC	0	2	5
ACB	0	3	4
BAC	2	0	5
BCA	1	0	6
CAB	4	3	0
CBA	1	6	0
	8	14	20

$$\varphi = \tfrac{1}{6}(8, 14, 20) = (1\tfrac{1}{3}, 2\tfrac{1}{3}, 3\tfrac{1}{3}),$$

which is the same answer we derived above. For an example of how to arrive at the numbers in the table, consider the order BCA. The value added by each player is calculated as follows:

$$\begin{aligned} &\text{B:} \quad v(\text{B}) - v(\phi) = 0 - 0 = 0 \\ &\text{C:} \quad v(\text{BC}) - v(\text{B}) = 6 - 0 = 6 \\ &\text{A:} \quad v(\text{ABC}) - v(\text{BC}) = 7 - 6 = 1. \end{aligned}$$

We can think of the Shapley value of player i in a game as the average amount that player i contributes when the grand coalition forms, given that all orders of coalition formation are equally likely. It would seem fair to give player i this average contribution. Of course, the real argument for the Shapley value as a fair imputation is based on Shapley's theorem.

The calculation of φ based on orders is conceptually simple, but for larger games, it becomes infeasible, since the number $n!$ of orders grows very rapidly as n increases. For example, consider the 4-person constant-sum game

$$v(A) = v(B) = v(C) = v(D) = 0$$
$$v(AB) = 50 \qquad v(CD) = 70 \qquad v(AC) = 30$$
$$v(BD) = 90 \qquad v(AD) = 30 \qquad v(BC) = 90$$
$$v(ABC) = v(ABD) = v(ACD) = v(BCD) = v(ABCD) = 120.$$

Game 26.2

The Shapley value, calculated in Table 26.1, is

$$\varphi = \tfrac{1}{24}(440, 920, 760, 760) = (18\tfrac{1}{3}, 38\tfrac{1}{3}, 31\tfrac{2}{3}, 31\tfrac{2}{3}).$$

Notice how B's strong bargaining position and A's weak one are mirrored in the Shapley value.

	Value added by			
Order	A	B	C	D
ABCD	0	50	70	0
ABDC	0	50	0	70
ACBD	0	90	30	0
ACDB	0	0	30	90
ADBC	0	90	0	30
ADCB	0	0	90	30
BACD	50	0	70	0
BADC	50	0	0	70
BCAD	30	0	90	0
BCDA	0	0	90	30
BDAC	30	0	0	90
BDCA	0	0	30	90
CABD	30	90	0	0
CADB	30	0	0	90
CBAD	30	90	0	0
CBDA	0	90	0	30
CDAB	50	0	0	70
CDBA	0	50	0	70
DABC	30	90	0	0
DACB	30	0	90	0
DBAC	30	90	0	0
DBCA	0	90	30	0
DCAB	50	0	70	0
DCBA	0	50	70	0
	440	920	760	760

Table 26.1: Shapley value calculation for Game 26.2.

This calculation makes it clear that we would not want to write out the table for calculating the Shapley value for a 5-person game! Fortunately, the calculation can be simplified by focussing on an individual player and asking how often he contributes how much to the forming grand coalition. When player i joins the forming grand coalition, he and the players who have already joined make up some coalition S, of size s, which contains player i. The amount i contributes is $v(S) - v(S - i)$. Furthermore, this contribution occurs for exactly those orderings in which i is preceded by the $s - 1$ other players in S, and followed by the $n - s$ players not in S. The number of orderings in which this happens is $(s - 1)!\,(n - s)!$. Hence we get the following expression for the Shapley value of player i:

$$(26.3)\quad \varphi_i = \frac{1}{n!}\sum_{i\in S}(s-1)!(n-s)!\,[v(S) - v(S-i)] \qquad (s = \text{the size of } S)$$

The summation is over all coalitions S which contain i. There are 2^{n-1} such coalitions, which is considerably fewer than the $n!$ terms we would add in the table calculation. For example, the calculation of φ_A in Game 26.2 could be done as follows:

Coalition S	$(s-1)!(n-s)!$	$v(S) - v(S-i)$	Product
A	$1 \times 6 = 6$	$0 - 0 = 0$	0
AB	$1 \times 2 = 2$	$50 - 0 = 50$	100
AC	$1 \times 2 = 2$	$30 - 0 = 30$	60
AD	$1 \times 2 = 2$	$30 - 0 = 30$	60
ABC	$2 \times 1 = 2$	$120 - 90 = 30$	60
ABD	$2 \times 1 = 2$	$120 - 90 = 30$	60
ACD	$2 \times 1 = 2$	$120 - 70 = 50$	100
ABCD	$6 \times 1 = 6$	$120 - 120 = 0$	0
			440

$$\varphi_A = \tfrac{1}{24}440 = 18\tfrac{1}{3}.$$

As a final observation, I should point out that the Shapley value is not the only candidate for an imputation which gives a fair allocation of payoffs. Indeed, we will see other candidates in Chapter 31, and compare them with the Shapley value in applications in Chapter 32. Since if we were sure that Shapley's axioms characterize fairness, the Shapley value *would* give the only fair allocation, at least some game theorists must question some of the axioms. Most people find Axioms 1 and 2 quite reasonable; discussion has focussed on Axiom 3. It has been argued, for instance by Luce and Raiffa [1957], that when players are playing two games v and w at once, there may be interaction effects between the games which would alter players' fair expectations in the games. However, it is certainly true that the Shapley value is by far the most important and widely used fair division method in game theory. The articles in [Roth, 1988] survey a variety of its uses.

Exercises for Chapter 26

1. For each of the following games, compute the Shapley value. Is it in the core?
   a) One seller, two buyers—Game (25.1).
   b) The Communications Satellite Game (25.2).
   c) The committee game from Chapter 19, Exercise 4a.
   d) The microcomputer game from Chapter 19, Exercise 4b. (It would be illuminating to do this both by a table, and by using formula 26.3.)
   e) The Glove Market, Chapter 25, Exercise 5. [Hint: if you use symmetry and call the players L, L, L, R, R, you need only consider ten distinct orderings.]
2. You can use the fact that the Shapley value for Divide-the- Dollar is $(\frac{1}{3}, \frac{1}{3}, \frac{1}{3})$ to find a fair division of payoffs for any game which is strategically equivalent to Divide-the-Dollar (see Chapter 23). Find a fair division of payoffs for Game 19.1.
3. For a general 3-person game in 0-normalized form

$$v(1) = v(2) = v(3) = 0$$
$$v(12) = a \qquad v(13) = b \qquad v(23) = c$$
$$v(123) = d$$

show that the Shapley values are given by

$$\varphi_1 = \frac{a+b+2(d-c)}{6} \qquad \varphi_2 = \frac{a+c+2(d-b)}{6} \qquad \varphi_3 = \frac{b+c+2(d-a)}{6}.$$

27. Application to Politics: The Shapley-Shubik Power Index

In a voting body, the voting rule specifies which subsets of players are large enough to pass bills, and which are not. Those subsets which can pass bills are called *winning coalitions*, while those which cannot are called *losing coalitions*. We can model a voting body as a characteristic function form game by assigning a value of 1 to all winning coalitions and 0 to all losing coalitions. The resulting game, in which all coalitions have a value of either 0 or 1, is called a *simple game*. A simple game is completely specified once we know its winning coalitions, and it is traditional to require it to satisfy several conditions which are reasonable for voting bodies.

DEFINITION. A *simple game* (or *voting game*) is a pair $(N, \mathcal{W})$, where N is the set of players (voters) and $\mathcal{W}$ is the collection of winning coalitions, such that

i) $\phi \notin \mathcal{W}$ (the empty set is a losing coalition)

ii) $N \in \mathcal{W}$ (the coalition of all voters is winning)

iii) $S \in \mathcal{W}$ and $S \subseteq T$ implies $T \in \mathcal{W}$ (if S is a winning coalition, so is any coalition which contains S).

A simple game is *proper* if $S \in \mathcal{W}$ implies $N - S \notin \mathcal{W}$, so that there cannot be two disjoint winning coalitions. It is *strong* if $S \notin \mathcal{W}$ implies $N - S \in \mathcal{W}$, so that for any coalition, either it or its complement is winning. Hence in a strong game voting deadlocks are not possible. Simple games representing voting bodies are almost always proper, but they are often not strong. For example, a voting body in which $\frac{2}{3}$ of the votes are necessary to pass a bill is not strong.

One common type of simple game is a *weighted voting game*, which is represented by the symbol

$$[q; w_1, w_2, \ldots, w_n].$$

Here there are n voters, w_i is the number of votes cast by voter i, and q is the *quota* of votes necessary to pass a bill. Thus a coalition S is winning if its total number of votes is at least q:

$$S \in \mathcal{W} \text{ if and only if } \sum_{i \in S} w_i \geqslant q.$$

Weighted voting games are important—we will see a number of examples—but not all important voting games can be represented as weighted voting games. For example, the United States legislative scheme is a simple game with 537 players: the President, 101 Senators (I am counting the Vice-President as a Senator since he can break tie votes in the Senate) and 435 Representatives. A coalition is winning if it contains

i) the President and a majority of both Senators and Representatives, or
ii) two-thirds of both Senators and Representatives (so it can override a Presidential veto).

Suppose we have modeled a voting body as a simple game, and we would like to measure the relative power of different voters in the game. For example, in the U.S. legislative scheme it seems clear that the President is more powerful than a Senator, who in turn is more powerful than a Representative, but could we quantify in some reasonable way how much more powerful? Even for the class of weighted voting games the problem is subtle. The power of voter i is certainly not always proportional to the number of votes w_i which i casts, as the following games illustrate:

[3; 2, 2, 1]	[4; 2, 2, 2, 1]	[3; 3, 1, 1].
A B C	A B C D	A B C
Game 27.1	Game 27.2	Game 27.3

In the first game the winning coalitions are AB, AC, BC, and ABC, and voter C plays a symmetric role with the other voters. Any reasonable measure of voting power should say that the voters have equal power. (My five year old once proposed this as a voting rule for our family. Since my wife and I are older, he was willing to allow each of us two votes!) In Game 27.2, on the other hand, voter D has no power at all. A bill will pass if two of A, B, or C vote for it, and fail if two of them vote against it, and how voter D votes never matters. Voter D is a *dummy* in this game. A reasonable measure of voting power should give him zero power. Finally, in Game 27.3 both B and C are dummies and A is a *dictator*: a bill will pass if and only if A votes for it. A power measure should give all the power to A.

In 1954 Lloyd Shapley and Martin Shubik pointed out that the Shapley value of a voting game could serve as a measure of voting power. Indeed, since φ is an imputation we have

$$\varphi_i \geqslant 0 \text{ for all voters } i,$$

$$\sum_{i \in N} \varphi_i = 1,$$

and Axioms 1 and 2 imply that

$$\varphi_i = \varphi_j \quad \text{if } i \text{ and } j \text{ play symmetric roles in the game,}$$

$$\varphi_i = 0 \quad \text{if } i \text{ is a dummy,}$$

$$\varphi_i = 1 \quad \text{if } i \text{ is a dictator.}$$

Hence the Shapley value gives the desired results for Games 27.1, 27.2, and 27.3. Furthermore, the procedure for calculating the Shapley value has a particularly nice interpretation for voting games. We imagine that a coalition forms in favor of some bill. The voters join one at a time, perhaps modifying the bill to gain the support of new voters. For a time, the coalition is losing, but at some point a voter i joins who makes the coalition winning, and by condition iii) for a simple game, it remains winning thereafter. The voter i who has this effect is said to be *pivotal* for this order of coalition formation, and she certainly has a good deal of power. The Shapley value of a voter is the proportion of orders in which that voter is pivotal, or equivalently, the probability that that voter will be pivotal if all orders of coalition formation are equally likely.

Shapley and Shubik's measure was widely adopted, and is known as the *Shapley-Shubik power index* of a voting game. Let us consider some examples. First, consider the weighted voting game

$$\begin{matrix} [6; & 4, & 3, & 2, & 1]. \\ & \text{A} & \text{B} & \text{C} & \text{D} \end{matrix}$$

Game 27.4

To calculate the Shapley-Shubik power index for this game, we can write out the 4! orderings of the voters and in each ordering underline the pivotal voter:

$$\begin{matrix} \text{A}\underline{\text{B}}\text{CD} & \text{B}\underline{\text{A}}\text{CD} & \text{C}\underline{\text{A}}\text{BD} & \text{DA}\underline{\text{B}}\text{C} \\ \text{A}\underline{\text{B}}\text{DC} & \text{B}\underline{\text{A}}\text{DC} & \text{C}\underline{\text{A}}\text{DB} & \text{DA}\underline{\text{C}}\text{B} \\ \text{A}\underline{\text{C}}\text{BD} & \text{BC}\underline{\text{A}}\text{D} & \text{CB}\underline{\text{A}}\text{D} & \text{DB}\underline{\text{A}}\text{C} \\ \text{A}\underline{\text{C}}\text{DB} & \text{BC}\underline{\text{D}}\text{A} & \text{CB}\underline{\text{D}}\text{A} & \text{DB}\underline{\text{C}}\text{A} \\ \text{AD}\underline{\text{B}}\text{C} & \text{BD}\underline{\text{A}}\text{C} & \text{CD}\underline{\text{A}}\text{B} & \text{DC}\underline{\text{A}}\text{B} \\ \text{AD}\underline{\text{C}}\text{B} & \text{BD}\underline{\text{C}}\text{A} & \text{CD}\underline{\text{B}}\text{A} & \text{DC}\underline{\text{B}}\text{A} \end{matrix}$$

Counting pivots gives $\varphi = (\frac{1}{24})(10, 6, 6, 2) = (\frac{5}{12}, \frac{3}{12}, \frac{3}{12}, \frac{1}{12})$. It is reasonable that B and C have the same power in this game even though they have different numbers of votes, since if we write out the winning coalitions—AB, AC, ABC, ABD, ACD, BCD, ABCD—we see that B and C do play symmetric roles in the game.

If we find writing out all the orderings too time-consuming, the Shapley value formula (26.3) yields a formula for the Shapley-Shubik power index of voter i:

$$\varphi_i = \frac{1}{n!} \sum_{i \text{ swings in } S} (s-1)!\,(n-s)! \qquad (s = |S|) \tag{27.5}$$

Here the sum is over all coalitions S such that $i \in S$, $S \in \mathcal{W}$, and $S - i \notin \mathcal{W}$ (so that $v(S) - v(S - i) = 1 - 0 = 1$). In words, the sum is over all coalitions S which are winning, but would become losing if i deserted the coalition. In such a situation, i is said to be a *swing voter* for S, and I have used the shorter phrase "i swings in S" to denote this. In the example above, voter D, for instance, is a swing voter only for S = BCD with $s = 3$, so

$$\varphi_D = \frac{1}{24}(2!\,1!) = \frac{2}{24} = \frac{1}{12}.$$

In many situations calculation of the Shapley-Shubik index can be simplified by taking advantage of symmetries in the voting game and using combinatorial techniques. As examples, we will consider two games analyzed in [Shapley and Shubik, 1954]. The first is the United Nations Security Council, which in 1954 consisted of five permanent members (the United States, the Soviet Union, Britain, France and China) and six non-permanent members. Resolutions had to be approved by seven members, but each of the five permanent members had veto power. Shapley and Shubik interpreted this to mean that a winning coalition had to have at least seven members, including all five of the permanent members. To calculate the Shapley-Shubik power index we will use the symmetry of the game and denote a permanent member by P and a non-permanent member by N. The total number of orderings of PPPPPNNNNNN is $11!/5!\,6! = 462$. The next step is to calculate in how many of these orderings an N is pivotal. An N will be pivotal precisely when it is preceded by all five P's and one other N, and followed by four N's:

$$(\text{PPPPPN})\underline{\text{N}}(\text{NNNN}).$$

The parentheses are meant to indicate that any possible arrangement of letters in parentheses will still result in the underlined N pivoting. The number of possible arrangements of PPPPPN is 6, while the number of possible arrangements of NNNN is 1. Hence the total number of arrangements in which an N pivots is $6 \times 1 = 6$.

Our conclusion is that the N's together have only 6/462 of the power, and the P's must have the other 456/462. Hence the power index of each individual P or N is

$$\varphi_P = \frac{1}{5} \cdot \frac{456}{462} = .1974$$

$$\varphi_N = \frac{1}{6} \cdot \frac{6}{462} = .0022.$$

It is quite startling that the six non-permanent members of the Security Council held, together, only 1.3% of the power in the Council, and that the ratio of the power of a permanent member to that of a non-permanent member was 91:1.

In 1965 the Security Council was expanded to include ten non-permanent members, with a total of nine members needed to pass a resolution (and each of the five permanent members still having a veto). If this was an attempt to increase the proportion of power held by non-permanent members, we can ask how successful it was. The Security Council is now PPPPPNNNNNNNNNN and the total number of orderings is $15!/5!\,10! = 3003$. Of these, an N will pivot only in the situation

$$(\text{PPPPPNNN})\underline{\text{N}}(\text{NNNNNN}).$$

The number of such orderings is $(8!/5!\,3!)(6!/6!) = 56$. Hence the N's now hold, together, $56/3003 = .0186$, or about 1.9% of the power, and the ratio of power of a permanent member to a non-permanent member is 105 : 1. The improvement has certainly been modest!

[Shapley and Shubik, 1954] also considered an example of a simple legislative scheme consisting of a president P, three senators S, and five representatives R. A bill had to be approved by the president and a majority of both the senators and the representatives. (They assumed for simplicity that there was no possibility of overriding a presidential veto.) In this game the players are PSSSRRRRR and the total number of orderings is $9!/1!\,3!\,5! = 504$. For an R to pivot it must be preceded by P, exactly two other R's, and either two or three S's. Here is the total number of orderings:

$$\begin{array}{ll r} \text{(PSSRR)}\underline{\text{R}}\text{(SRR):} & (5!/1!\,2!\,2!)(3!/1!\,2!) = & 90 \\ \text{(PSSSRR)}\underline{\text{R}}\text{(RR):} & (6!/1!\,3!\,2!)(2!/2!) = & \underline{60} \\ & & 150 \end{array}$$

Similarly, for an S to pivot it must be preceded by P, exactly one other S, and either three, four or five R's:

$$\begin{array}{ll r} \text{(PSRRR)}\underline{\text{S}}\text{(SRR):} & (5!/1!\,1!\,3!)(3!/1!\,2!) = & 60 \\ \text{(PSRRRR)}\underline{\text{S}}\text{(SR):} & (6!/1!\,1!\,4!)(2!/1!\,1!) = & 60 \\ \text{(PSRRRRR)}\underline{\text{S}}\text{(S):} & (7!/1!\,1!\,5!)(1!/1!) = & \underline{42} \\ & & 162 \end{array}$$

In the remaining $504 - 150 - 162 = 192$ orderings, P pivots. The division of power is as follows:

President	$192/504 = .381$		
Total senators	$162/504 = .321$	Each senator	.107
Total representatives	$150/504 = .298$	Each representative	.060

The ratio of power of P : S : R is 192 : 54 : 30, or 6.4 : 1.8 : 1.

If one does this calculation for the United States legislative scheme with 101 S's and 435 R's, the result is that the President would have, if no veto override were possible, almost exactly half the voting power, and the Senate and the House of Representatives would split the other half almost equally (the Senate having slightly more). The ratio of power of P : S : R would be 870 : 4.3 : 1. The inclusion of the possibility of Congress overriding a Presidential veto changes the picture dramatically. The power of P drops to about 1/6, with the remaining 5/6 split almost equally between the Senate and the House. The individual power ratios are 174 : 4.3 : 1.

For our last example, we will consider a legislative example from [Straffin, 1977a] which illustrates how the Shapley-Shubik power index might be used to study the dynamics of political coalition formation. The County Board of Rock County, Wisconsin, has 40 members and votes by majority rule. There are two cities in the county: Janesville with 14 board members and Beloit with 11 board members. The remaining 15 members come from small villages and rural areas. On some issues there has been considerable rivalry between Beloit and Janesville. However, the board members from these cities have been an independent lot, and no pattern of bloc voting has emerged among the members from either Beloit or Janesville, although at times Beloit officials have suggested that more "coherence" in voting by Beloit board members might be effective in promoting the city's interests. Our goal is to use the Shapley-Shubik power index to analyze the effect that bloc voting would have on the distribution of power in the County Board.

If there is no bloc voting, the game is

$$[21;\ \underbrace{1, 1, \ldots, 1}_{40 \text{ voters}}]$$

and each board member has $1/40$ of the power. Hence the Beloit members have, collectively, $11/40 = .275$, and the Janesville members have $14/40 = .350$.

Now suppose the board members from Beloit organize and agree to vote as a bloc. The resulting game is

$$[21;\ \underset{\text{B}}{11},\ \underbrace{1, \ldots, 1}_{29 \text{ others}}].$$

For this 30-player game there are $30!/1!\,29! = 30$ orderings. Voter B will pivot if he is preceded by from 10 to 20 other voters, i.e. in 11 of the 30 orderings. Hence Beloit has $11/30 = .367$ of the power. The Janesville board members have collectively $(14/29)(19/30) = .306$. The organized Beloit delegation has indeed gained power, at the expense of the unorganized Janesville delegation and the rural members.

Similarly the Janesville delegation would gain power if it organized and voted as a bloc while the Beloit members remained individualistic. However, the interesting case to consider is when *both* Beloit and Janesville members form voting blocs. After all, if members from one city organized, there should be considerable pressure on members from the other city to do likewise. With two voting blocs we have the game

$$[21;\ \underset{\text{B}}{11},\ \underset{\text{J}}{14},\ \underbrace{1, \ldots, 1}_{15 \text{ others}}].$$

The number of orderings is $17!/1!\,1!\,15! = 272$. It is easiest to see when a "1" would pivot. It would have to be preceded by B and exactly nine other 1's, or by J and exactly six other 1's:

$$\begin{array}{lr} (\text{B}111111111)\underline{1}(\text{J}11111): & (10!/1!\,9!)(6!/1!\,5!) = 60 \\ (\text{J}111111)\underline{1}(\text{B}11111111): & (7!/1!\,6!)(9!/1!\,8!) = \underline{63} \\ & 123 \end{array}$$

It is a little harder to calculate the number of orderings in which B would pivot. B would have to be preceded by J and from zero to six 1's, or preceded by from ten to all fifteen ones (hence *followed* by J and from zero to five 1's). The number of ways in which these things could happen is

$$1 + 2 + 3 + 4 + 5 + 6 + 7 = 28 \quad \text{plus} \quad 1 + 2 + 3 + 4 + 5 + 6 = 21$$

for a total of 49 orderings. Thus B has $49/272 = .180$ of the power. J must have $(272 - 123 - 49)/272 = .368$.

To help interpret these results, we can place them in a 2×2 game matrix.

		Janesville		
		Doesn't organize		Organizes
Beloit	Doesn't organize	(.275, .350)	⟶	(.204, .519)
		↓		↑
	Organizes	(.367, .306)	⟶	(.180, .368)

The movement diagram for this non-constant-sum game shows that Janesville has a dominant strategy of organizing, and that if Janesville organizes, Beloit is actually better off not organizing. The political conclusion might be that if the Janesville board members are not engaging in bloc voting, from a power perspective the Beloit members should certainly not provoke them to do so by organizing first. The broader point is that the effects of bloc voting in legislatures on the power of groups of legislators can be subtle, and the Shapley-Shubik index is a useful tool in analyzing such shifts of power.

The Shapley-Shubik power index has been used in an enormous number of contexts. Indeed, [Shapley and Shubik, 1954] was one of the dozen most cited papers in social science literature in the period 1955–1980. [Lucas, 1983] and [Lambert, 1989] survey applications in European parliaments, New York county boards and the United States Electoral College, and also give useful information on computational methods. [Brams, 1976] discusses a number of interesting paradoxes of voting power. [Straffin, 1983] gives additional dynamic applications to bloc formation, quarreling legislators and bandwagons. It also contains, in an appendix, a listing of all distinct simple games with four or fewer players, together with their Shapley-Shubik indices.

Exercises for Chapter 27

1. List the winning coalitions in each of the following games. Is the game proper? Is it strong?
   a) $[7; 4, 3, 2, 1]$
   b) $[5; 3, 2, 1, 1]$
   c) Five voters A, B, C, D, E vote by majority rule, but A has a veto.
   d) Voters A, B, C, D sit in order. To win, a coalition must include two voters who sit next to each other.
2. Calculate Shapley-Shubik power indices for the games in 1. Take advantage of symmetry where you can.
3. What would be the effect on power if the 1965 Security Council required a resolution to be approved by ten members instead of nine (keeping the permanent members' veto)?

4. What would be the effect on power if the 1-3-5 legislative system were modified to include the possibility of Congress overriding a presidential veto by $\frac{2}{3}$ vote in both houses (i.e. 2-1 in the Senate, 4-1 in the House)?

5. a) Suppose a 9-member legislature has a 3-member committee. To pass, a bill must get a majority in the committee and a majority in the full legislature. (Assume that a committee member who votes for the bill in committee will also vote for it on the house floor.) What is the ratio of power of a committee member to a legislator who is not on the committee?
   b) Suppose the committee is enlarged to 5 members. What is the effect on power? (The United States House of Representatives not long ago enlarged its Rules Committee as part of legislative reform.)

6. Allen, Bates, Cowley, Doolittle, Evans, Faraday and Gale are going to form a joint stock company. Bates, who put up most of the capital, will have 41% of the stock, with the rest being split equally among the other six partners. At the last minute, it is proposed to admit Harris, letting Bates keep his 41% but now splitting the rest seven ways instead of six.

   Bates objects strongly. Allen can't understand what difference it should make to Bates. Cowley, who knows some game theory, figures that Bates should be pleased, since diluting the rest of the holdings should increase Bates' power.

   Illuminate this situation by calculating Bates' power a) if Harris is not admitted, b) if Harris is admitted.

7. The County Board of Nassau County, New York in the 1950's and '60's had six members, and operated by weighted voting since the members represented towns of different sizes. The table gives the numbers of votes members cast in 1958 and 1964.

District	Votes in 1958	Votes in 1964
Hempstead #1	9	31
Hempstead #2	9	31
North Hempstead	7	21
Oyster Bay	3	28
Glen Cove	1	2
Long Beach	1	2
Needed to pass:	16	58

   Calculate the Shapley-Shubik power indices for the board in 1958 and in 1964. In particular, why did John Banzhaf [1965] sue, claiming voters in some towns were not adequately represented?

8. My student Pauline Parkinson [1981] studied the distribution of power in the black student organization Jinyosha (the word is swahili for "harmony") at Beloit College in 1980. The club had 30 members, 19 men and 11 women, and required a $\frac{2}{3}$ vote to pass resolutions. There was an active group of 8 women who formed a voting bloc, and an active group of 12 men who considered forming a rival bloc. Use the Shapley-Shubik power index to analyze the effects of these potential blocs on voting power.

9. Find a voting body of your choice and do a Shapley-Shubik power analysis of it.

28. Application to Politics: The Banzhaf Index and the Canadian Constitution

When John Banzhaf wished to demonstrate the inequity of the Nassau County Board (Chapter 27, Exercise 7), he did not use the Shapley-Shubik index, but devised his own power index. Banzhaf reasoned that a voter only has a direct effect on the voting outcome when he is a swing voter in some winning coalition, and hence a voter's power should be proportional to the number of coalitions in which that voter is a swing voter. The resulting index is known as the *Banzhaf index*, and it is sometimes used as an alternative to the Shapley-Shubik index.

To see how the Banzhaf index works, let us calculate it for some of the examples in Chapter 27. Consider Game 27.4:

$$\begin{matrix}[6; & 4, & 3, & 2, & 1]. \\ & \text{A} & \text{B} & \text{C} & \text{D}\end{matrix}$$

To calculate the Banzhaf index we write out not the orderings of the voters, but the winning coalitions. Then in each winning coalition we underline the swing voters, those whose defection would make the coalition losing:

$$\underline{\text{AB}} \quad \underline{\text{AC}} \quad \underline{\text{A}}\text{BC} \quad \underline{\text{AB}}\text{D} \quad \underline{\text{AC}}\text{D} \quad \underline{\text{BCD}} \quad \text{ABCD}$$

In coalition ABC, for example, A is a swing voter since BC is not winning, but B is not a swing voter because AC is winning. Now count the number of swings for each voter, and divide by the total number of swings for all voters (to make the indices add to one). The result is the Banzhaf power index

$$\beta = (\tfrac{5}{12}, \tfrac{3}{12}, \tfrac{3}{12}, \tfrac{1}{12}).$$

For this particular game, it turns out that the Banzhaf index is identical to the Shapley-Shubik index. You can check that this is also true for Games 27.1, 27.2, and 27.3. However, it is not always true. The simplest game for which the two indices differ is

$$\begin{matrix}[3; & 2, & 1, & 1]. \\ & \text{A} & \text{B} & \text{C}\end{matrix}$$

Game 28.1

Here it is easy to check that the Shapley-Shubik index is $\varphi = (\frac{2}{3}, \frac{1}{6}, \frac{1}{6})$. However, for the Banzhaf index we get swings $\underline{\text{AB}}$, $\underline{\text{AC}}$, $\underline{\text{A}}\text{BC}$, and $\beta = (\frac{3}{5}, \frac{1}{5}, \frac{1}{5})$. Thus the ratio of power of A to B is 4 : 1 by the Shapley-Shubik index, but only 3 : 1 by the Banzhaf index. For very small games the differences between the two indices are minor, but for larger games they can be significant. Let us consider some other examples from Chapter 27. In doing so, we will also see how combinatorial techniques are used in calculating the Banzhaf index.

First, the 1954 U.N. Security Council: PPPPPNNNNNN. We need to write down all *types* of winning coalitions with swing voters underlined, calculate in how many ways each type can appear, and hence calculate the total number of swings for P's and N's. The calculation looks like this:

Swings for Type	Number of ways	P's	N's
PPPPPNN	$\binom{5}{5}\binom{6}{2} = 1 \cdot 15 = 15$	75	30
PPPPPNNN	$\binom{5}{5}\binom{6}{3} = 1 \cdot 20 = 20$	100	0
PPPPPNNNN	$\binom{5}{5}\binom{6}{4} = 1 \cdot 15 = 15$	75	0
PPPPPNNNNN	$\binom{5}{5}\binom{6}{5} = 1 \cdot 6 = 6$	30	0
PPPPPNNNNNN	$\binom{5}{5}\binom{6}{6} = 1 \cdot 1 = 1$	5	0
		285	30

$$\beta_P = (1/5)(285/315) = .1810 \qquad \beta_N = (1/6)(30/315) = .0159.$$

The "number of ways" calculation is done by noting that in PPPPPNN, for example, we have selected all five P's, but only two N's from a possible six N's. The number of ways of selecting two objects from six objects is the binomial coefficient $\binom{6}{2} = 6!/2!\,(6-2)! = 15$.

The non-permanent members have $30/315 = .095$ of the swings, hence 9.5% of the Banzhaf power, which is about seven times as much power as they had by the Shapley-Shubik index. The power ratio of a permanent to a non-permanent member is 11 : 1 by the Banzhaf index, as compared to 91 : 1 by the Shapley-Shubik index.

Next consider Shapley and Shubik's legislature PSSSRRRRR from Chapter 27. Here is the Banzhaf calculation:

Swings for Type	Number of ways	P	S's	R's
PSSRRR	$\binom{1}{1}\binom{3}{2}\binom{5}{3} = 1 \cdot 3 \cdot 10 = 30$	30	60	90
PSSRRRR	$\binom{1}{1}\binom{3}{2}\binom{5}{4} = 1 \cdot 3 \cdot 5 = 15$	15	30	0
PSSRRRRR	$\binom{1}{1}\binom{3}{2}\binom{5}{5} = 1 \cdot 3 \cdot 1 = 3$	3	6	0
PSSSRRR	$\binom{1}{1}\binom{3}{3}\binom{5}{3} = 1 \cdot 1 \cdot 10 = 10$	10	0	30
PSSSRRRR	$\binom{1}{1}\binom{3}{3}\binom{5}{4} = 1 \cdot 1 \cdot 5 = 5$	5	0	0
PSSSRRRRR	$\binom{1}{1}\binom{3}{3}\binom{5}{5} = 1 \cdot 1 \cdot 1 = 1$	1	0	0
		64	96	120

The division of power is

President	64/280 = .229		
Total senators	96/280 = .343	Each senator	.114
Total representatives	120/280 = .429	Each representative	.086

The ratio of power of P : S : R is 64 : 32 : 24 or 8 : 4 : 3. This is quite different from the Shapley-Shubik result. When the calculation is done for the United States legislative scheme the difference is even more serious. According to the Banzhaf index, with no veto override the President has less than 4% of the power, the Senate 31% and the House of Representatives 65%. The addition of the possibility of a veto override *does not change* these figures appreciably!

For a final example, we will consider an impressive scheme for amending the Canadian constitution, proposed at the Victoria Conference in 1971. The problem in designing a constitutional amendment scheme for Canada is that the ten Canadian provinces are very jealous of their constitutional prerogatives, and extraordinarily diverse both in politics and in size. The provinces of Ontario and Quebec together contained 64% of the Canadian population in 1970, whereas the four small "Atlantic" provinces together contained less than 10%. This extreme diversity certainly suggests asymmetric treatment of the provinces in a constitutional amendment scheme, but exactly how to do it is a delicate matter. The Victoria scheme proposed that a constitutional amendment would have to be approved by

i) both Ontario and Quebec, and
ii) at least two of the four Atlantic provinces (New Brunswick, Nova Scotia, Newfoundland and Prince Edward Island), and
iii) British Columbia and at least one prairie province (Alberta, Saskatchewan, Manitoba) *or* all three prairie provinces.

Table 28.1 shows the Shapley-Shubik and Banzhaf power indices for this scheme, together with the relative populations of the provinces in 1970. (You can try your hand at verifying some of these figures in Exercise 4.) Two things are remarkable from the table. The first is how very well the Shapley-Shubik index matches the percentages of population. Although the scheme was not designed with knowledge of the index, its designers certainly had an accurate feeling for relative power by Shapley and Shubik's measure. The second observation is what different results the Banzhaf index gives: Ontario and Quebec are much less powerful, and the smaller provinces are much more powerful. Indeed, this was the first published example in which the two indices differed in the *order* of power among the players. The Shapley-Shubik index says a prairie province is more powerful than an Atlantic province and the Banzhaf index says the reverse.

By the way, the constitutional amendment scheme which Canada finally adopted in 1982, after fifty-five years of debate and provincial negotiations, was far less equitable than the Victoria scheme, in the sense that voting power as measured by either index does not approximate population at all. For example, Ontario's proportion of power is 14.4% by Shapley-Shubik and 12.3% by Banzhaf, while Prince Edward Island's is 8.7% by Shapley-Shubik and 9.3% by Banzhaf. The complete story of negotiations is told from a power perspective in [Kilgour and Levesque, 1984].

Province	Percentage of power: Shapley-Shubik index φ	Percentage of power: Banzhaf index β	Percentage of population (1970)	
Ontario	31.55	21.78	34.85	average 31.90
Quebec	31.55	21.78	28.94	
British Columbia	12.50	16.34	9.38	
Alberta	4.17	5.45	7.33	average 5.65
Saskatchewan	4.17	5.45	4.79	
Manitoba	4.17	5.45	4.82	
New Brunswick	2.98	5.94	3.09	average 2.47
Nova Scotia	2.98	5.94	3.79	
P.E.I.	2.98	5.94	0.54	
Newfoundland	2.98	5.94	2.47	

Table 28.1: Values of Power Indices for a proposed Canadian Constitutional Amendment Scheme, from [Straffin, 1977c].

What are we to make of examples like these? We would like to have a convincing way to measure power in voting bodies. The Shapley-Shubik index, nicely specialized from the axiomatically elegant Shapley value of game theory, seems perfect for the job. However, since 1965 there have been two power indices, both defined in seemingly reasonable ways and yet giving measures that are sometimes significantly, even startlingly, different. Which should we believe?

To a mathematician, the clearest way to lay bare the logical nature of a construction is to characterize it axiomatically. We could write down axioms which characterize the two indices and then compare those axioms to see which set is most likely to apply in a given political situation. [Dubey, 1975] and [Dubey and Shapley, 1979] did this very elegantly, adapting Shapley's axioms for the value to the specific context of simple games to characterize the Shapley-Shubik index, and then changing one of the axioms to get a characterization of a form of the Banzhaf index.

Another approach is to derive both indices from a common model, with one difference in assumptions leading to the difference in the resulting measures. In [Straffin 1977c, 1983, 1988], both indices are viewed as answers to the natural question "How likely is it that voter i's ballot will actually make a difference between whether a bill passes or fails?" The difference between the two indices comes from different assumptions about how voters' votes are related to each other. The assumption that there is no relation—that voter i's vote has no effect on how voter j votes—leads to a form of the Banzhaf index. The Shapley-Shubik index arises from the assumption that voters' votes are correlated with each other in a way which can be interpreted as saying that all voters judge bills according to some common set of standards.

This approach would have us ask, in any given application, whether there is reason to believe that such a common set of standards might exist. If so, use the Shapley-Shubik index. On the other hand, if voters appear to be completely independent, use the Banzhaf index. For example, in the U.S. legislative scheme I would argue that it is reasonable to assume some common standards, and the Shapley-Shubik index is more applicable. Indeed, the extremely unintuitive results of the Banzhaf index (the President with less than 4% of the power and veto overrides having negligible effect on power) might well arise from the extreme inappropriateness in this situation of the assumption that the President and all members of Congress vote independently of each other. On the other hand, given the ideological diversity of the Canadian provinces, the Banzhaf index might better measure power relations in that situation.

One last caution is in order. In the discussion of power in Chapter 27 and this chapter, it should be very clear that we are talking only about *voting* power. There are certainly many other determinants of power than just a legislator's place in a voting scheme: knowledge, rational persuasion, ideological or emotional appeal, threats, logrolling and all the richness of politics. All of these are abstracted away in our measures of voting power. That isn't all bad. One can argue that at the constitutional level at which voting schemes are designed, most of these other considerations should be abstracted away. What is left, moreover, still has a surprising amount of richness.

Exercises for Chapter 28

1. Calculate Banzhaf indices for the games in Chapter 27, Exercise 1.
2. Calculate the Banzhaf indices for the 1965 U.N. Security Council (see Chapter 27).
3. What are the Banzhaf indices for the Nassau County Board in 1958 and 1964 (see Chapter 27, Exercise 7)? This is the situation which provoked Banzhaf's winning lawsuit.
4. Verify the calculation of the power of an Atlantic province under the Victoria scheme. Let the players be OOAAAABPPP. Use
   a) the Shapley-Shubik index. [Hint: you should find that the A's pivot in 1500 out of 12600 orderings.]
   b) the Banzhaf index. [Hint: the total number of swings is 404. You should find that A's swing in four different types of coalitions, for a total of 96 swings.]

29. Bargaining Sets

Consider the game in characteristic function form

$$v(A) = v(B) = v(C) = 0$$
$$v(AB) = 60 \qquad v(AC) = 80 \qquad v(BC) = 100$$
$$v(ABC) = 105$$

Game 29.1

If we plot the core conditions as in Figure 29.1, we see that the core is empty. This game is not constant-sum, so not equivalent to Divide-the-Dollar, and its von Neumann-Morgenstern stable sets are complicated. The Shapley value is easy to compute (Exercise 1), but is not meant to describe an outcome which might actually occur if this game were played. Furthermore, the solution theories for non-constant-sum games which we have considered so far assume that the outcome of this game will be an imputation, i.e. that the players will divide the full 105 units. However, it is possible to imagine that coalition $\{BC\}$, say, would form, leaving A out and dividing only 100 units. Group rationality would argue that $\{BC\}$ should admit A and divide the extra five units, but in practice players don't always do that, perhaps because the added 5 units might be felt not to be worth the effort of extra negotiation.

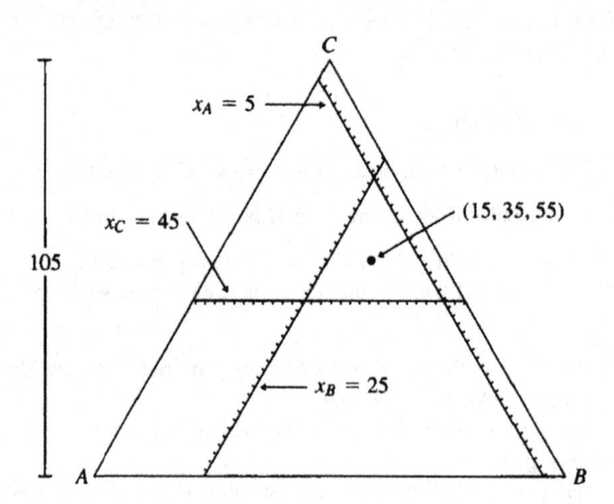

Figure 29.1: The empty core and the bargaining set for $\{ABC\}$ for Game 29.1

[Aumann and Maschler, 1964] proposed a solution concept for this kind of situation which is now known as the theory of *bargaining sets*. (Actually, there are several versions of bargaining sets, but we will consider only the simplest one.) The first idea is that we will not prejudice the discussion of which coalition will form by assuming that it will be the grand coalition. In fact, we will consider every possible coalition structure, and for each structure we will try to see what division of payoffs might result from bargaining among the players. For Game 29.1 there are five possible coalition structures:

$$\{A\}\{B\}\{C\} \qquad \{AB\}\{C\} \qquad \{AC\}\{B\} \qquad \{A\}\{BC\} \qquad \{ABC\}.$$

Some of the resulting payoffs are clear. For the coalition structure $\{A\}\{B\}\{C\}$ in which no player is willing to cooperate with any other, all players will end up with 0. By the way, this can happen. We all know situations in which bargaining breaks down and everyone goes away empty-handed. Similarly, in the coalition structure $\{A\}\{BC\}$, A should get 0, and B and C should split 100, so the result would not be an imputation. We need to think about how B and C should split the 100. Suppose the payoffs to B and C are x_B and x_C, with $x_B, x_C \geqslant 0$ and $x_B + x_C = 100$. Would the bargaining positions of the players put additional restrictions on x_B and x_C?

Let us imagine that B and C consider the proposed division $(x_B, x_C) = (50, 50)$. This doesn't seem to reflect the situation of the game, in which C has a stronger position than B, since coalition $\{AC\}$ has greater value than coalition $\{AB\}$. Aumann and Maschler imagine C objecting to this proposed division, based on what C could get if she formed a coalition with A.

C's objection against B: "If you insist on (50, 50), I can form a coalition with A. Since $v(AC) = 80$, I could offer A 25 and keep 55 myself. A would agree to my proposal since he is currently getting 0, and I would be better off than I currently am with you."

We have seen that two can play the game of defection, so B could try meeting C's objection by a counter-objection.

B's attempted counter-objection against C: "If you attempt to form a coalition with A, I could compete with you. I would match your offer of 25 to A, and since $v(AB) = 60$ I could keep 35 for myself, which is..."

At this point, B's voice trails off into embarrassed silence, because 35 is *less* than the 50 he is currently claiming. C has an objection which cannot be met by a valid counter-objection. The inference is that C is getting too little in the proposed division of payoff.

Now suppose that B and C consider a division of (40, 60). C, emboldened by her previous success, objects that she still isn't getting enough.

C's objection against B: "I could form a coalition with A, give him 15, and keep 65 for myself. A would agree, and I would get more than my 60 under the current proposal."

Now B can fight.

B's counter-objection against C: "If you try that, I can compete by offering A up to 20, while keeping at least the 40 I am currently getting."

In fact, B can compete with any amount C could offer A while keeping $x_C > 60$. Similarly, C can compete with any amount B could offer A while keeping $x_B > 40$. We say the proposal of $(40, 60)$ is *stable* for $\{BC\}$, since any objection by one player against another can be met by a valid counter-objection.

The condition for stability is easily found. For proposal (x_B, x_C), C's maximal offer to A is $v(AC) - x_C = 80 - x_C$, while B's maximal offer is $v(AB) - x_B = 60 - x_B$. The condition for stability is that these maximal offers be equal:

$$60 - x_B = 80 - x_C, \qquad x_C - x_B = 20.$$

Combining this with $x_B + x_C = 100$, we can solve for the unique division $(x_B, x_C) = (40, 60)$. This is the bargaining set division of payoff for the coalition $\{BC\}$.

Aumann and Maschler's idea works for any coalition structure in any game. Here's how the formalization looks. For a game in characteristic function form (N, v), a *coalition structure* is a partition $N = S_1 \cup S_2 \cup \cdots \cup S_k$, where the S_j's are disjoint coalitions. A payoff n-tuple $\mathbf{x} = (x_1, \ldots, x_n)$ is *rational* for this coalition structure if

$x_i \geqslant v(i)$	for all i	(it is individually rational), and
$\sum_{i \in S_j} x_i = v(S_j)$	for all j	(each coalition in the structure divides its full value)

Notice that in general $(x_1, \ldots, x_n)$ is not an imputation.

Now consider two players I and J in the same coalition S_j. (We don't allow players in different coalitions to object against each other.) I has an *objection* against J if there is some coalition $S \subseteq N$ and payoff n-tuple $\mathbf{y} = (y_1, \ldots, y_n)$ such that

Ob1) S contains I but not J	(I threatens to form S and leave out J)
Ob2) $y_k > x_k$ for all $k \in S$	(all members of S prefer $\mathbf{y}$ to $\mathbf{x}$)
Ob3) $\sum_{k \in S} y_k = v(S)$	(S can assure its members what is proposed in $\mathbf{y}$)

Notice that S will not be one of the original S_j's. Player I is threatening to disrupt the current coalition structure.

J has a valid *counter-objection* against I if there is some other coalition T and payoff n-tuple $\mathbf{z} = (z_1, \ldots, z_n)$ such that

COb1) T contains J but not I	
COb2) $\sum_{k \in T} z_k = v(T)$	(T can assure its members what is proposed in $\mathbf{z}$)
COb3) $z_k \geqslant x_k$ for all $k \in T$	(all members of T like $\mathbf{z}$ at least as much as $\mathbf{x}$)
COb4) $z_k \geqslant y_k$ for all $k \in T \cap S$	($\mathbf{z}$ is competitive with $\mathbf{y}$)

We say $\mathbf{x}$ is *stable* for the coalition structure $S_1, \ldots, S_k$ if every objection can be met by a valid counter-objection. The set of all stable rational n-tuples for the coalition structure is the *Aumann-Maschler bargaining set* for that structure.

The bargaining sets for all the coalition structures in Game 29.1 turn out to be unique n-tuples. They are

Coalition structure	A	B	C
$\{A\}\{B\}\{C\}$	0	0	0
$\{AB\}\{C\}$	20	40	0
$\{AC\}\{B\}$	20	0	60
$\{A\}\{BC\}$	0	40	60
$\{ABC\}$	15	35	55

Notice the balancing property of the bargaining sets for the two-player coalitions, and that the bargaining set for the grand coalition preserves the same differences between the players' payoffs. If you plot the bargaining set for $\{ABC\}$ in Figure 29.1, it is the center of the small triangle formed by the core constraints.

Something interesting happens if we modify the game by increasing the value of the grand coalition:

$$v(A) = v(B) = v(C) = 0$$

$$v(AB) = 60 \qquad v(AC) = 80 \qquad v(BC) = 100$$

$$v(ABC) = 135$$

Game 29.2

Since we have not changed the values of the two-player coalitions, and all objections and counter-objections in two-player coalitions involve only two-player coalitions, the bargaining sets for two-player coalitions are the same as for Game 29.1. However, Figure 29.2 shows that this game has a non-empty core. If the grand coalition $\{ABC\}$ is considering any payoff n-tuple in the core, the core conditions tell us that no player can have an objection, since conditions

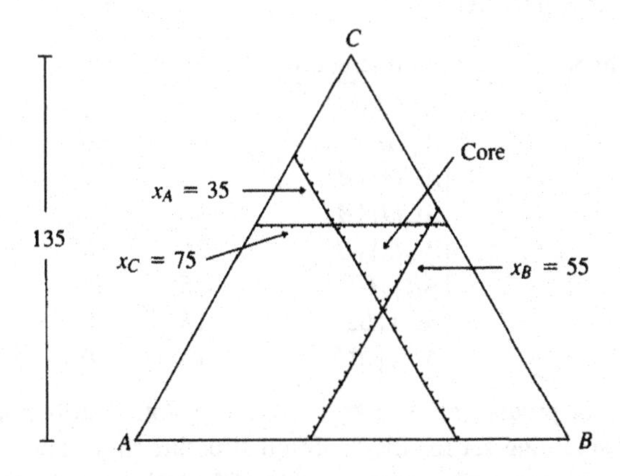

Figure 29.2: The core of Game 29.2.

Ob2 and Ob3 cannot both be met. Hence the entire core is in the bargaining set for coalition structure $\{ABC\}$. In fact, the bargaining set for $\{ABC\}$ is the core (Exercise 3).

This example shows that the bargaining set for a coalition structure may be more than a single n-tuple. Aumann and Maschler were able to prove, though, that the bargaining set is never the empty set. Any coalition structure for a game in characteristic function form must have at least one stable rational payoff n-tuple. (In Chapter 30 we will see that the bargaining set may be empty for some coalition structures in other kinds of games, with interesting consequences.)

The example also illustrates that payoff divisions in the bargaining set may not always seem "fair." For example, A might well not be pleased with $(5, 55, 75)$ for the coalition structure $\{ABC\}$. However, A cannot "object" by finding another coalition in which both he and his partner would do better. He *could* threaten to desert the three-person coalition, which would have some force, since then B and C could only split 100. However, this kind of protest is not considered in the Aumann-Maschler theory. We will consider it in another solution concept in Chapter 31.

In my Game Theory class, groups of four students played a modified version of Game 26.2:

$$v(A) = v(B) = v(C) = v(D) = 0$$

$$v(AB) = 50 \qquad v(CD) = 70 \qquad v(AC) = 30$$

$$v(BD) = 90 \qquad v(AD) = 30 \qquad v(BC) = 90$$

$$v(ABC) = v(ABD) = v(ACD) = v(BCD) = 120$$

$$v(ABCD) = 0$$

Game 29.3

I set $v(ABCD) = 0$ to destroy the superadditivity of the game and force the players to join small coalitions, in order to test the bargaining set idea. The results for eight groups were

Group number	Coalition structure	A	B	C	D
1	$\{ABC\}\{D\}$	26	42	52	0
2*	$\{A\}\{BCD\}$	0	50	35	35
3*	$\{ACD\}\{B\}$	30	0	45	45
4*	$\{ACD\}\{B\}$	34	0	43	43
5*	$\{ACD\}\{B\}$	26	0	47	47
6*	$\{AB\}\{CD\}$	25	25	35	35
7	$\{AC\}\{BD\}$	15	30	15	60
8	$\{AD\}\{BC\}$	15	40	50	15

Five of the eight groups (marked by *) chose payoff distributions in or very close to the bargaining set for the coalition structure they chose. In particular, the three groups with coalition structure $\{ACD\}\{B\}$ averaged the bargaining set distribution $(30, 0, 45, 45)$ (see Exercise 5). In groups 1, 7, and 8 player B has objections against C or D which cannot be met by valid counter-objections (Exercise 6), and hence is settling for a payoff which is "too small." The reason for this is illuminating. Players in this game quickly noticed B's strong bargaining position, and there was a tendency for the weaker players to gang up on the strong player and form coalition structure $\{ACD\}\{B\}$. To prevent this from happening, B had to sacrifice the strength of his bargaining position.

One conclusion is that social and psychological factors—for instance the tendency in democratic United States society for the weak to seek solidarity against the strong—can and do modify purely rational considerations when people bargain. Given this, I think the bargaining set idea did an impressive job of predicting outcomes in this game. In the next chapter, we will see a context in which bargaining sets have not worked as well.

Exercises for Chapter 29

1. Calculate the Shapley values for Games 29.1 and 29.2. How do they compare with the bargaining set for the grand coalition?
2. Check that for Game 29.1 the bargaining set for the grand coalition is indeed the single imputation $(15, 35, 55)$. For instance, for the imputation $(10, 35, 60)$ find an objection which cannot be met by a valid counter-objection.
3. Check that for Game 29.2, the bargaining set for the grand coalition is indeed the core, by choosing some imputations not in the core, and finding objections which cannot be met by valid counter-objections. For example, try $(15, 65, 55)$ and $(40, 60, 35)$.
4. Find the bargaining sets for the game
$$v(A) = v(B) = v(C) = 0 \quad v(AB) = 5 \quad v(AC) = 8 \quad v(BC) = 9 \quad v(ABC) = 10.$$
5. Do some checks toward showing that in Game 29.3 the bargaining set for the coalition structure $\{ACD\}\{B\}$ is the single imputation $(30, 0, 45, 45)$:
   a) For the imputation $(40, 0, 40, 40)$, find an objection which cannot be met by a valid counter-objection.
   b) For $(30, 0, 45, 45)$, suppose A objects against C by proposing $\{ABD\}$ and payoff division $(31, 43, 0, 46)$. Find a valid counter-objection.
   c) For $(30, 0, 45, 45)$, suppose C objects against A by proposing $\{BC\}$ and payoff division $(*, 44, 46, *)$ (we don't care what the $*$'s are). Find a valid counter-objection.
6. For each of the coalition structures and payoff divisions 1, 7, and 8 in my class experiment, show that B has an objection which cannot be met by a valid counter-objection.

30. Application to Politics: Parliamentary Coalitions

In a parliamentary democracy with more than two major parties, it is common that no single party will have a majority of seats in the parliament. Hence a majority government must be formed by a coalition of parties. In this chapter we will consider several game theoretic approaches to the study of parliamentary coalitions.

As an example, consider the 1965 parliamentary election in Norway, which gave results

Party	Number of seats
A. Labor	68
B. Christian	13
C. Liberal	18
D. Center	18
E. Conservative	31
	148

It takes 75 members to form a coalition government. Can we predict which parties formed the government?

In the weighted voting game

$$\begin{matrix}[75; & 68, & 13, & 18, & 18, & 31] \\ & A & B & C & D & E\end{matrix}$$

there are five coalitions which are *minimal winning*, in the sense that they are winning, but would not be winning if any player were omitted. These are *AB*, *AC*, *AD*, *AE*, and *BCDE*. William Riker [1962] argued that since it would not be advantageous to divide the spoils of governing (cabinet posts, patronage appointments, implementation of favored programs, etc.) among more parties than are necessary, we could predict that the governing coalition which forms should be minimal winning. This prediction is known in the political science literature as the *Riker size principle*. (Recall the Pathans of Chapter 24.) Although there are certainly exceptions, Riker's size principle is fairly well supported by political data, and is a good starting point for a theory of political coalitions.

In general, however, a game has many minimal winning coalitions. If we want to make a more specific prediction about *which* minimal winning coalition will form, there are at least two ways we could extend Riker's idea. First, suppose that the spoils of governing are divided in proportion to the *number of votes* which the parties in the governing coalition bring to the coalition, or at least that the parties perceive that this is how the spoils will be divided. (There is some evidence from experimental games for this perception.) Thus if coalition *AB* forms and governs,

A receives 68/81 of the spoils, and *B* gets 13/81. Alternatively, if *AC* forms, *A* gets 68/86 and *C* gets 18/86. Notice that *A* prefers the larger fraction 68/81 to the smaller 68/86, and hence would rather form a coalition with *B* than with *C*. In general, parties want to belong to a winning coalition with *as few votes* as possible, since this maximizes their share of the spoils. In our example

Minimal winning coalition:	*AB*	*AC*	*AD*	*AE*	*BCDE*
Number of votes:	81	86	86	99	80

we would predict that *BCDE* is the most likely coalition to form, with *AB* as second most likely.

On the other hand, suppose that the parties believe that spoils will be divided *equally* among coalition members. After all, since a minimal winning coalition becomes losing if any member defects, one can argue that all members are equally important. In this case, parties would maximize their share of the spoils by belonging to a winning coalition with *as few members* as possible. We would predict that *AB*, *AC*, *AD*, or *AE* would all be more likely than *BCDE*.

Alas, these two extensions of Riker's principle contradict each other in our example, and in fact neither one of them works very well in practice. This shouldn't be very surprising, since while making rather delicate assumptions about the division of spoils, we have left out of consideration something which any student of politics knows is important in the formation of political alliances: ideology. The Radicals and the Reactionaries might be a minimal-vote or minimal-member winning coalition, but we would not expect them to join easily. We need a way to take ideology into account.

The most common model of political ideology, dating back to the period immediately following the French Revolution, is to place politicians or political parties on a one-dimensional continuum from left (radical, liberal) to right (conservative, reactionary). We did this informally for Carter, Anderson and Reagan in Chapter 20. Here is a left-right placement, developed in [Converse and Valen, 1971] from survey data, for the five Norwegian parties in 1965, with respect to their stands on economic issues:

liberal	−5	0	4	6	7	11	conservative
	A		*B*	*C*	*D*	*E*	

The scale is an *interval scale* (Chapter 9), and could be altered by any linear function.

Robert Axelrod [1970], considering this kind of one-dimensional spatial placement, suggested that the governing coalitions which form should be *connected*, in the sense that they should contain all parties in some interval of the line. In this example the minimal winning connected coalitions are *AB* and *BCDE*. *AC*, for example, is not connected, since it does not include *B*, which is contained in any interval containing *A* and *C*. The prediction of a connected minimal winning coalition has a fair degree of empirical support. For example, [DeSwaan, 1973] found that more than half of the 108 coalition governments he studied were minimal winning and connected.

A serious problem with this one-dimensional spatial model is that it is often difficult to reduce complex ideological positions to placement on a one-dimensional continuum. What do you do, for example, with a party which is liberal on social issues, but conservative on economic issues? As an example, cultural issues are important to Norwegian voters, and here is Converse and Valen's placement of Norwegian parties by their stands on cultural issues:

liberal	1	2	3	6	10	conservative
	C	E	A	D	B	

Now the connected minimal winning coalitions are AE and AD.

The most natural mathematical way out of this difficulty is to plot ideological positions as points in a Cartesian plane, where the first coordinate represents economic issues and the second coordinate represents cultural issues. If there were three salient kinds of issues, we could use a three dimensional plot. Figure 30.1 shows the two-dimensional plot of Norwegian parties in 1965. In a two-dimensional plot, it is customary to measure the relative ideological proximity of parties by the standard Euclidean distance. In Figure 30.1, for example,

$$d(A,B) = \sqrt{(-5-4)^2 + (3-10)^2} = \sqrt{130} \approx 11.4$$
$$d(A,C) = \sqrt{(-5-6)^2 + (3-1)^2} = \sqrt{125} \approx 11.2.$$

Thus C is slightly closer to A than B is. When we measure distances like this, of course, we are making quite strong assumptions about the comparability of scales for economic issues and cultural issues.

If we have parties represented as points in an n-dimensional ideological space, with each party weighted by its number of seats in parliament, we can try to use game-theoretic reasoning to predict which parties will form a governing

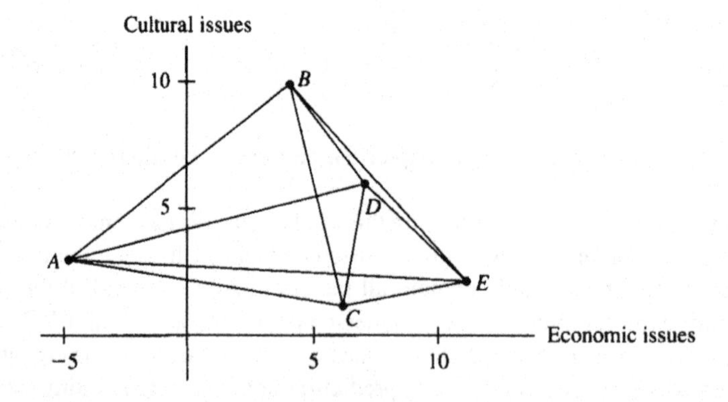

Figure 30.1: Two-dimensional spatial placement of Norwegian parties

coalition. Since the Aumann-Maschler bargaining sets are most directly tied to coalitional structure, it would seem most natural to use that approach. However, in a spatial game with ideologically concerned parties, the most natural object of bargaining is not "division of spoils." Instead, the parties bargain about what kinds of policies the government they form will pursue. These *platforms* can also be represented as points in ideological space. Each party would like to be part of a coalition which adopts a platform close to that party's ideological point. It is thus points in ideological space, rather than n-tuples of payoffs, which are the basis of offers, objections and counter-objections.

To understand how the bargaining set idea works in this context, let us consider a simpler example than the Norwegian parliament. Figure 30.2 shows five parties F, G, H, J, and K, whose ideological points are $(0,0)$, $(0,10)$, $(5,15)$, $(10,10)$, and $(10,0)$ respectively. We will suppose they are all of comparable size, so that any three of them form a majority. Suppose coalition $\{GHJ\}$ considers platform $\mathbf{a} = (5,10)$, which does look geometrically reasonable for that coalition (see Figure 30.2a). Unfortunately, this platform is not stable in Aumann and Maschler's sense.

G objects against H: "I can go to F and K and propose platform $\mathbf{b} = (3,7)$. This is better than $\mathbf{a}$ for all of us in $\{FGK\}$."

You should check G's claim, remembering that "better" means "closer to my position." To this objection, H has no valid counter-objection. A counter-objection

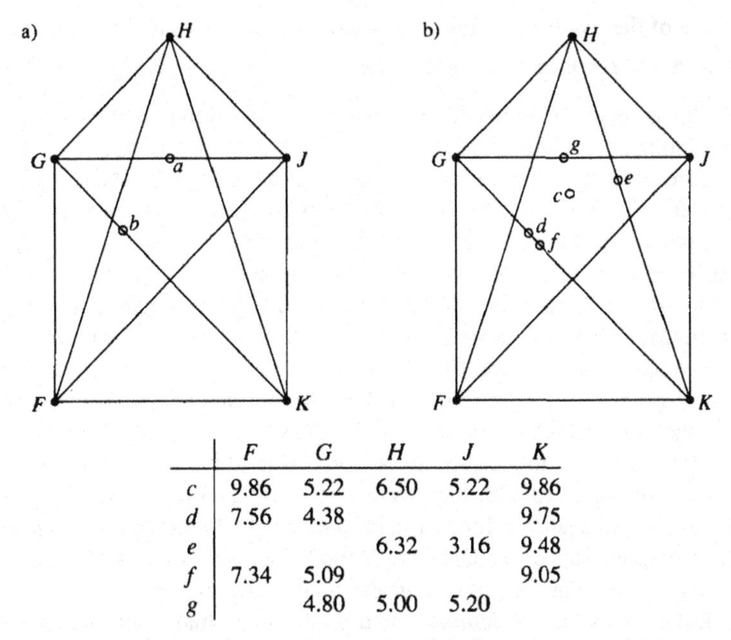

	F	G	H	J	K
c	9.86	5.22	6.50	5.22	9.86
d	7.56	4.38			9.75
e			6.32	3.16	9.48
f	7.34	5.09			9.05
g		4.80	5.00	5.20	

Figure 30.2: Hypothetical parties, proposed platforms, and some distances

would require a platform which is at least as close as **a** to H, and no such platform could be as close as **b** to either F or K.

In fact, it turns out that coalition $\{GHJ\}$ has *no* stable platform: the bargaining set for $\{GHJ\}$ is empty. However, there are winning coalitions which do have stable platforms. Consider coalition $\{FHK\}$ with platform $\mathbf{c} = (5, 8.5)$ (see Figure 30.2b). There are possible objections to this platform, but they can all be met by valid counter-objections. For example,

F objects against H: "I can go to G and K and propose platform $\mathbf{d} = (3.1, 6.9)$. This is better than **c** for all of us in $\{FGK\}$."

H counter-objects against F: "If you try that, I can go to J and K and propose platform $\mathbf{e} = (7, 9)$. This is better than **c** was for J and me, and it is better than **d** for K."

You should use the table of distances in Figure 30.2b to check these claims by F and H. Exercise 2 asks you to consider some other possible objections and find counter-objections for them.

In this game the five "external coalitions" $\{FGH\}$, $\{GHJ\}$, $\{HJK\}$, $\{FJK\}$, and $\{FGK\}$ do not have stable platforms, and the five "internal coalitions" $\{FHK\}$, $\{GJK\}$, $\{FHJ\}$, $\{GHK\}$, and $\{FGJ\}$ do have stable platforms. (If you locate these coalitions in Figure 30.2 I think you'll see what the words "external" and "internal" refer to.) Hence Aumann-Maschler bargaining theory would predict that

i) one of the five internal minimal winning coalitions should form, and
ii) it should propose a platform in the central pentagon of Figure 30.2.

I like the Aumann-Mascher objection/counter-objection reasoning, but at least in this example the conclusions it gives seem suspect. It says that in a situation like Figure 30.2 the coalition which forms should be one which "splits the opposition." For example, two conservative parties might join with one liberal party, leaving out two moderate parties. The platform of the governing coalition should be centrist, appealing just as much to the parties left out of the coalition as to the parties included. That just doesn't seem to be what happens!

In fact, [McKelvey, Ordeshook, and Winer, 1978] ran some experiments based on a game like that in Figure 30.2. Five players were given a diagram showing their ideal points, and were told that they should choose, by majority rule, a point in the plane as their "outcome." Each player would be paid an amount of money depending on how close the outcome was to his ideal point. Since the players could win up to $20, they were well motivated to bargain. In all eight trials an *external* coalition formed, and *none* of the outcomes was within the central pentagon. McKelvey, Ordeshook and Winer were motivated to reject the bargaining set solution and propose an alternative solution for spatial games (they called it the *competitive solution*) which predicts external coalitions instead of internal ones. See their paper, or [Straffin and Grofman, 1984].

In the example of Figure 30.1 the bargaining set theory predicts that *AB*, *AC*, *AD*, or *AE* will form, while McKelvey, Ordeshook, and Winer's competitive solution predicts *AB*, *AC*, or *BCDE*. The coalition which actually formed was *BCDE*. (Recall that *BCDE* is also predicted by Riker's non-ideological theory when spoils are divided in proportion to votes.)

The spatial model we have used to study ideological coalitions has also been used quite effectively to study voting behavior, and is an active area of research in political science. [Enelow and Hinich, 1984] and [Straffin, 1989b] survey some of the results. The Shapley-Shubik power index of Chapter 27 has been adapted by Shapley [1977] to this spatial voting context. [Rabinowitz and Macdonald,1986] gives an interesting application of the spatial power index to the United States electoral college.

Exercises for Chapter 30

1. a) Show that $[75; 68, 13, 18, 18, 31]$ is equivalent to $[4; 3, 1, 1, 1, 1]$.
   b) Compute the Shapley-Shubik power index for this game.

2. Do some further checks that $\mathbf{c}$ is a stable platform for $\{FHK\}$. Use the platforms $\mathbf{d}, \mathbf{e}, \mathbf{f} = (3.6, 6.4)$ and $\mathbf{g} = (4.8, 10)$ and the table of distances in Figure 30.2b. For each of the following objections, show that it is an objection, and find a valid counter-objection.
   a) F objects against H, proposing $\{FGK\}$ with $\mathbf{f}$.
   b) H objects against F, proposing $\{HGJ\}$ with $\mathbf{g}$.
   c) H objects against F, proposing $\{HJK\}$ with $\mathbf{e}$.

31. The Nucleolus and the Gately Point

The wheel that squeaks the loudest
Is the one that gets the grease.
—Josh Billings

The Shapley value of an n-person game in characteristic function form is a single imputation which might be proposed as a "solution" to the game. It is based on axioms embodying a concept of fairness, rather than on bargaining considerations. In this chapter we will look at two single-imputation solutions based on concepts of bargaining.

The first, proposed by David Schmeidler [1970], is known as the *nucleolus*. Recall that the core consists of all imputations $\mathbf{x} = (x_1, \ldots, x_n)$ which satisfy

$$\sum_{i \in S} x_i \geqslant v(S) \qquad \text{for every } S \subseteq N.$$

If the core is empty, no imputation satisfies all of these constraints. However, we could try to satisfy them as nearly as possible. One way to interpret this would be to make the largest violation as small as possible, and this is Schmeidler's idea. For every imputation $\mathbf{x}$ and coalition $S \subseteq N$, define the *excess* of S at $\mathbf{x}$, by

$$e_S(\mathbf{x}) = v(S) - \sum_{i \in S} x_i.$$

Thus $e_S(\mathbf{x})$ is the difference between what the members of S could get if they bestirred themselves and exerted the power of S, and what they are getting in the imputation $\mathbf{x}$. We could think of it as a measure of the unhappiness of S with $\mathbf{x}$. Our goal is to find the imputation or imputations $\mathbf{x}$ which will minimize the largest of the $e_S(\mathbf{x})$'s. In other words, we try to make the most unhappy coalition as little unhappy as possible. In Josh Billings' terms, we grease the squeakiest wheel, until the most squeaking wheel squeaks as little as possible.

Let us see how this works for Game 29.1:

$$v(A) = v(B) = v(C) = 0$$
$$v(AB) = 60 \qquad v(AC) = 80 \qquad v(BC) = 100$$
$$v(ABC) = 105$$

Suppose we start with an arbitrary imputation, say $\mathbf{x} = (20, 35, 50)$. We compute

$$\begin{aligned}
e_A(\mathbf{x}) &= 0 - 20 = -20 \\
e_B(\mathbf{x}) &= 0 - 35 = -35 \\
e_C(\mathbf{x}) &= 0 - 50 = -50 \\
e_{AB}(\mathbf{x}) &= 60 - (20 + 35) = 5 \\
e_{AC}(\mathbf{x}) &= 80 - (20 + 50) = 10 \\
e_{BC}(\mathbf{x}) &= 100 - (35 + 50) = 15 \\
e_{ABC}(\mathbf{x}) &= 105 - (20 + 35 + 50) = 0.
\end{aligned}$$

Notice that the conditions for an imputation mean that the excess of a one-person coalition will always be negative or zero, and the excess of the grand coalition will always be zero.

The largest excess is $e_{BC}(\mathbf{x}) = 15$. We can lower that by choosing an imputation which gives more to coalition BC, hence less to A. Since $e_{AC}(\mathbf{x})$ is more than $e_{AB}(\mathbf{x})$, it would seem to be a good idea to take 5 from A and give it to C. Hence we try $\mathbf{y} = (15, 35, 55)$, and recompute the excesses of the two-player coalitions.

$$\begin{aligned}
e_{AB}(\mathbf{y}) &= 60 - 50 = 10 \\
e_{AC}(\mathbf{y}) &= 80 - 70 = 10 \\
e_{BC}(\mathbf{y}) &= 100 - 90 = 10.
\end{aligned}$$

Clearly we have reached the minimum, since lowering any one of these excesses would raise another one. Interpreted geometrically, we have found the centroid of the triangle of core constraints, the point whose maximal distance from these constraints is as small as possible. See Figure 31.1. This point is called the *nucleolus* of the game, and we will denote it by the Greek letter ν. Thus, for Game 29.1, $\nu = (15, 35, 55)$.

It sometimes happens that there are several different imputations which minimize the largest excess. To choose among them, Schmeidler's natural idea was

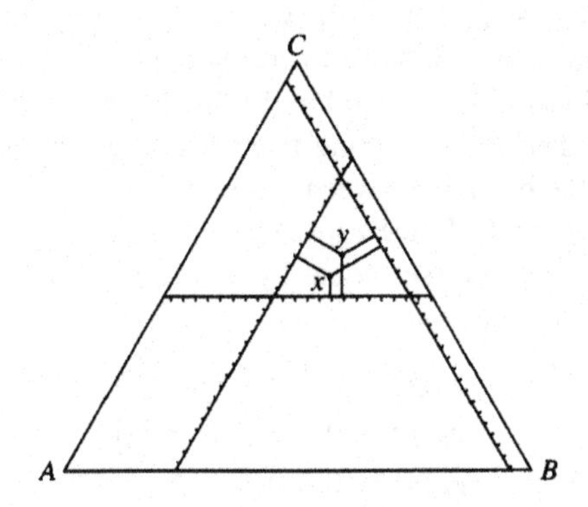

Figure 31.1: Finding the nucleolus of Game 29.1.

to minimize the second largest excess, then the third largest, and so on until we come down to a unique imputation, which is then the nucleolus.

We should also consider the case when the core is non-empty. In this case, for any imputation $\mathbf{x}$ in the core, all of the excesses $e_S(\mathbf{x})$ are zero or negative. If we follow Schmeidler's recipe of minimizing the largest excess, we will find the imputation in the core which makes the least negative excess as negative as possible. Seen geometrically this is the point in the core whose distance from the closest wall of the core is as large as possible. In other words, we will find the point which is as far inside the core as possible. Thus the nucleolus seems to be a very reasonable candidate for a solution in the case of a non-empty core as well.

As an example of these ideas, consider the game

$$v(A) = v(B) = v(C) = 0$$
$$v(AB) = 4 \qquad v(AC) = 0 \qquad v(BC) = 3$$
$$v(ABC) = 6$$

Game 31.1

and start with the imputation $\mathbf{x} = (x_A, x_B, x_C) = (2, 3, 1)$. Then

$$\begin{aligned} e_A(\mathbf{x}) &= 0 - 2 = -2 \\ e_B(\mathbf{x}) &= 0 - 3 = -3 \\ e_C(\mathbf{x}) &= 0 - 1 = -1 \\ e_{AB}(\mathbf{x}) &= 4 - 5 = -1 \\ e_{AC}(\mathbf{x}) &= 0 - 3 = -3 \\ e_{BC}(\mathbf{x}) &= 3 - 4 = -1 \\ e_{ABC}(\mathbf{x}) &= 6 - 6 = 0. \end{aligned}$$

The largest excess is $e_{ABC}(\mathbf{x}) = 0$. We cannot lower this, since the excess of the grand coalition is zero for any imputation. Hence we proceed to the next largest excesses $e_C(\mathbf{x}) = e_{AB}(\mathbf{x}) = e_{BC}(\mathbf{x}) = -1$. Notice that we cannot lower the excesses of either C or AB, since lowering one would raise the other, so we must leave these alone as the second and third largest excesses. However, we can lower the excess of BC by taking from A and giving to BC. Since we do not want to disturb the balance between C and AB, we should take some amount ϵ from A and give it to B. Since $e_{BC}(\mathbf{x}) = -1$ and $e_A(\mathbf{x}) = -2$, the best we can do is to make these excesses equal by taking $\epsilon = 1/2$. Thus we are led to $\mathbf{y} = (1.5, 3.5, 1)$, with

$$\begin{aligned} e_A(\mathbf{y}) &= 0 - 1.5 = -1.5 \\ e_B(\mathbf{y}) &= 0 - 3.5 = -3.5 \\ e_C(\mathbf{y}) &= 0 - 1 = -1 \\ e_{AB}(\mathbf{y}) &= 4 - 5 = -1 \end{aligned}$$

$$e_{AC}(\mathbf{y}) = 0 - 2.5 = -2.5$$
$$e_{BC}(\mathbf{y}) = 3 - 4.5 = -1.5$$
$$e_{ABC}(\mathbf{y}) = 6 - 6 = 0.$$

$\mathbf{y}$ is the nucleolus. In Figure 31.2 we see that it is the central point in the parallelogram-shaped core. In fact, at least for three-person games, it is easier to find the nucleolus by geometry than by guided trial and error as we did. If you know something about linear programming, the nucleolus of a large game can be found most efficiently by running a series of linear programs.

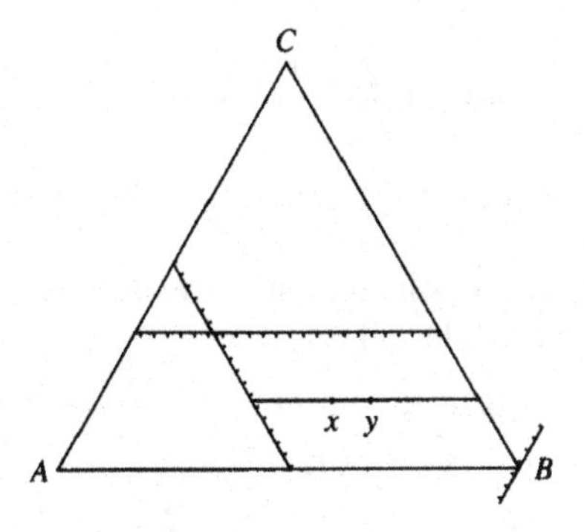

Figure 31.2: Finding the nucleolus in Game 31.1.

The nucleolus has a number of nice properties. It always exists and is a unique imputation. It is fairly easy to calculate. Unlike the Shapley value, it is always in the core if the core is non-empty, and it is always in the Aumann-Maschler bargaining set for the grand coalition. On the other hand, you will see in the exercises that the nucleolus can seem quite ruthless in comparison with the Shapley value.

One justification for considering the nucleolus as a solution, at least when the core is empty, comes from considerations of stability. To get an imputation as the payoff n-tuple, we need to have the grand coalition: all players must cooperate. If some subset S of players has too large an excess, they would be strongly tempted to go off and do better for themselves, thereby breaking up the grand coalition. To have the best hope of keeping the grand coalition together, we should make the most unhappy subset S as little unhappy as possible.

This kind of reasoning—considering the temptation to disrupt the grand coalition—was used explicitly as the basis for a different solution offered by Dermot Gately [1974a]. Consider a three-player game with players 1, 2, and 3 considering an imputation $\mathbf{x} = (x_1, x_2, x_3)$ as a proposed distribution of payoffs. Look at the situation from player 1's point of view. Player 1 is necessary to the grand coalition, and if she deserts, players 2 and 3 will lose an amount

$x_2 + x_3 - v(23)$, the difference between what they would get in $\mathbf{x}$ and what they could get without player 1. Of course, player 1 would also lose something, namely $x_1 - v(1)$. Gately defines player 1's *propensity to disrupt* the grand coalition to be the ratio of these amounts:

$$d_1(\mathbf{x}) = \frac{x_2 + x_3 - v(23)}{x_1 - v(1)}.$$

If $d_1(\mathbf{x})$ is large, player 1 can say, "If I desert the grand coalition, I'll lose something, but you'll lose a lot more!"

For a general n-person game, player i's propensity to disrupt an imputation $\mathbf{x}$ is defined by

$$d_i(\mathbf{x}) = \frac{\sum_{j \neq i} x_j - v(N - i)}{x_i - v(i)}.$$

The *Gately point* of a game is the imputation which minimizes the maximum propensity to disrupt.

Let us compute the Gately point for our two examples. First Game 31.1. Start with $\mathbf{x} = (x_A, x_B, x_C) = (2, 3, 1)$ and compute

$$d_A(\mathbf{x}) = \frac{x_B + x_C - v(BC)}{x_A - v(A)} = \frac{3 + 1 - 3}{2 - 0} = \frac{1}{2}$$

$$d_B(\mathbf{x}) = \frac{x_A + x_C - v(AC)}{x_B - v(B)} = \frac{2 + 1 - 0}{3 - 0} = 1$$

$$d_C(\mathbf{x}) = \frac{x_A + x_B - v(AB)}{x_C - v(C)} = \frac{2 + 3 - 4}{1 - 0} = 1.$$

We can lower d_B and d_C by taking something from A and giving to B and C. It turns out (I'll show you below how to find this) that we should take $4/11$ from A and give $3/11$ to B, $1/11$ to C. In other words, we should consider the new imputation $\mathbf{y} = (18/11, 36/11, 12/11)$:

$$d_A(\mathbf{y}) = \frac{\frac{36}{11} + \frac{12}{11} - 3}{\frac{18}{11} - 0} = \frac{5}{6}$$

$$d_B(\mathbf{y}) = \frac{\frac{18}{11} + \frac{12}{11} - 0}{\frac{36}{11} - 0} = \frac{5}{6}$$

$$d_C(\mathbf{y}) = \frac{\frac{18}{11} + \frac{36}{11} - 4}{\frac{12}{11} - 0} = \frac{5}{6}.$$

Since the propensities to disrupt are equal, and lowering any one means raising another one, this is the best we can do. Hence $\mathbf{y}$ is the Gately point.

This observation that the way to minimize the largest propensity to disrupt is to make all of the propensities to disrupt equal holds in general, and leads to a simple way to calculate the Gately point. Notice that

$$d_i(\mathbf{x}) = \frac{\sum_{j \neq i} x_j - v(N-i)}{x_i - v(i)} = \frac{v(N) - x_i - v(N-i)}{x_i} = \frac{v(N) - v(N-i)}{x_i} - 1,$$

if we assume that the game has been normalized so that $v(i) = 0$ for all i, and use the fact that $\sum_{i \in N} x_i = v(N)$ for an imputation. Hence the way to set all the $d_i(\mathbf{x})$ equal is to choose x_i in proportion to $v(N) - v(N-i)$. For Game 31.1 these proportions are $6-3:6-0:6-4$, or $3:6:2$. Hence the Gately point is $(3/11, 6/11, 2/11) \times 6 = (18/11, 36/11, 12/11)$, as we saw above.

If we apply this method to Game 29.1, we obtain the proportions $105-100:105-80:105-60$, or $1:5:9$. Hence the Gately point is $(1/15, 5/15, 9/15) \times 105 = (7, 35, 63)$. You might like to check that this imputation does equalize the three players' propensities to disrupt.

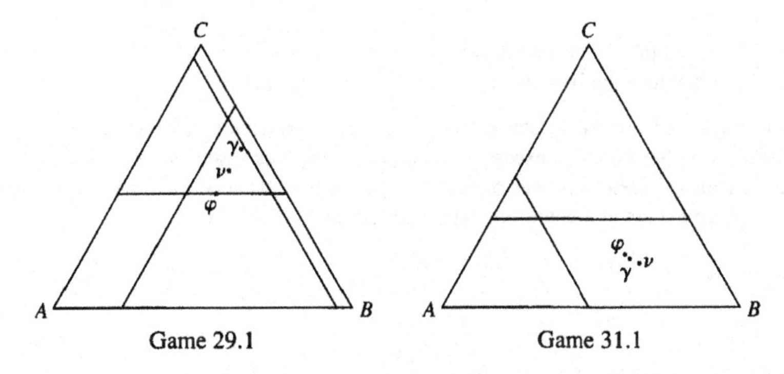

Figure 31.3: Comparing the nucleolus, Gately point, and Shapley value.

Figure 31.3 shows the core, Shapley value (φ), nucleolus (ν), and Gately point (γ) for Games 29.1 and 31.1. You can get a better feel for how these solutions relate by drawing plots like this for other examples in the exercises. Exercise 4 shows that the Gately point need not be in the core when the core is non-empty, if $n > 3$. However, Gately's idea can be extended by considering the propensities to disrupt of coalitions as well as individuals, and minimizing the largest of these in the same manner as for the nucleolus. Littlechild and Vaidya [1976] call the resulting imputation the *disruption nucleolus*, and it is always in any non-empty core.

Exercises for Chapter 31

1. Compute and compare the nucleolus, the Gately point and the Shapley value for each of the following games. I would recommend marking them in the imputation triangle, and also drawing the core constraints.

   a) Game 29.2 b) Game 26.1

2. Compute the nucleolus of the following weighted voting games, and compare it to the Shapley value (i.e. the Shapley-Shubik power index).
   a) [6;4,3,2,1] (Game 27.4) b) [7;4,3,2,1] (Exercise 27.1a)
   Do you think the nucleolus would make a good power index?

3. What is the nucleolus of
   a) One seller, two buyers (Game 25.1)?
   b) The glove market (Exercise 25.5)?

4. Consider the game

$$v(A) = v(B) = v(C) = v(D) = 0$$
$$v(AB) = 2 \qquad v(AC) = 3 \qquad v(AD) = 4$$
$$v(CD) = 7 \qquad v(BD) = 6 \qquad v(BC) = 5$$
$$v(ABC) = 5 \quad v(ABD) = 6 \quad v(ACD) = 7 \quad v(BCD) = 8$$
$$v(ABCD) = 10.$$

   a) Show that the core is non-empty.
   b) Compute the Gately point and show it is not in the core.

5. The nucleolus minimizes the largest excess. But perhaps it isn't as bad for a large coalition to have a large excess as it is for a small coalition to have a large excess. Investigate the idea of the *per capita nucleolus*, where what is minimized is the largest *per capita* excess of a coalition, $e_S(\mathbf{x})/|S|$. [Grotte, 1971/72]

32. Application to Economics: Cost Allocation in India

In this chapter we will consider a problem of cooperative hydroelectric power development in southern India in the 1970's. The analysis is due to Dermot Gately [1974a]. The players were the four states comprising the Southern Electricity Region of India: Tamil Nadu, Andhra Pradesh, Kerala, and Mysore. To simplify calculations and because of economic and geographical similarities, Gately considered Kerala and Mysore as a single state. He then used an integer programming model to predict costs of expanding and operating the electric power system in the region under five scenarios: each state acting individually, three possible pairs of two states cooperating, and all three states cooperating. The results are given in Table 32.1.

	Costs incurred in			
Coalition structure	T	A	K	Total
$\{T\}\{A\}\{K\}$	533	187	86	806
$\{T,K\}\{A\}$	260	187	242	689
$\{A,K\}\{T\}$	533	48	148	729
$\{T,A\}\{K\}$	552	147	86	785
$\{T,A,K\}$	301	101	251	653

T = Tamil Nadu $\quad$ A = Andhra Pradesh $\quad$ K = Kerala-Mysore
Units are 10 million rupees, value in 1974.

Table 32.1: Hydroelectric costs in south India.

We can see immediately that cooperation of all three states is economically most efficient, saving 153 units (1.53 billion rupees) from the cost of individual developments. The problem is to decide how the 653 unit cost of the joint project should be allocated among the states. If each state simply paid the cost incurred in that state, Kerala-Mysore would pay 251 units, which is considerably more than the 86 units it would pay if it refused to cooperate and developed its power individually. The same problem occurs if costs are apportioned proportionally to power usage: Kerala-Mysore was projected to use 31% of the power of the joint project, which would result in a cost of $.31 \times 653 = 202$ units, again more than its go-it-alone cost.

Let us do a game theoretic analysis. First we set up the "cost game":

$$C(T) = 533 \qquad C(A) = 187 \qquad C(K) = 86$$

$$C(TK) = 260 + 242 = 502$$

$$C(AK) = 48 + 148 = 196$$

$$C(TA) = 552 + 147 = 699$$

$$C(TAK) = 653.$$

We are interested in how the states might take advantage of the savings from a joint project. One way to analyze this is to subtract each coalition's cost from the sum of the individual costs of its members to get the "savings game" with $v(S) = \sum_{i \in S} C(i) - C(S)$:

$$v(T) = v(A) = v(K) = 0$$

$$v(TK) = 533 + 86 - 502 = 117$$

$$v(AK) = 187 + 86 - 196 = 77$$

$$v(TA) = 533 + 187 - 699 = 21$$

$$v(TAK) = 533 + 187 + 86 - 653 = 153.$$

Game 32.1

Figure 32.1 shows that the core of this game is non-empty. We would certainly like to recommend a savings allocation in the core. However, we would like to have a specific recommendation about how savings should be allocated. First,

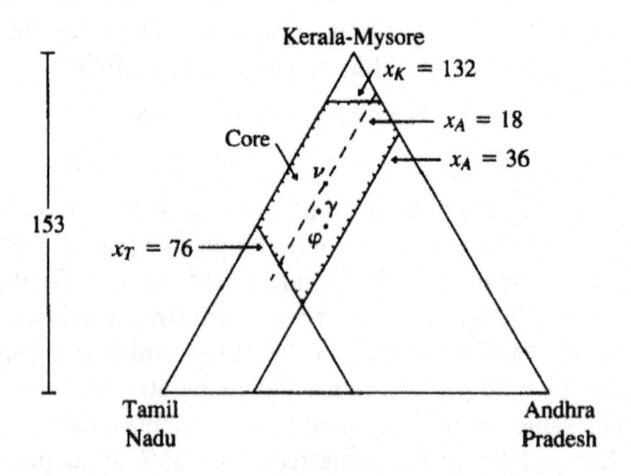

Figure 32.1: The core, nucleolus, Gately point, and Shapley value of Game 32.1.

we will calculate the Shapley value, using formula 26.3.

$$\begin{aligned}\varphi_T &= \frac{1}{3!}[2(v(T) - v(\phi)) + 1(v(TK) - v(K)) + 1(v(TA) - v(A)) \\ &\quad + 2(v(TAK) - v(AK))] \\ &= \tfrac{1}{6}[2 \times 0 + 1 \times 117 + 1 \times 21 + 2(153 - 77)] = 48\tfrac{1}{3}.\end{aligned}$$

and similarly

$$\begin{aligned}\varphi_A &= \tfrac{1}{6}[2 \times 0 + 1 \times 77 + 1 \times 21 + 2(153 - 117)] = 28\tfrac{1}{3} \\ \varphi_K &= \tfrac{1}{6}[2 \times 0 + 1 \times 117 + 1 \times 77 + 2(153 - 21)] = 76\tfrac{1}{3}.\end{aligned}$$

The imputation $\varphi = (48\frac{1}{3}, 28\frac{1}{3}, 76\frac{1}{3})$ does lie in the core.

Next, let us calculate the nucleolus. It is clear from Figure 32.1 that the nucleolus must lie on the line $x_A = 18$. On this line, it must lie halfway between the lines $x_K = 132$ and $x_T = 76$. Now if $x_A = 18$ and $x_T = 76$, then $x_K = 153 - 18 - 76 = 59$. Halfway between $x_K = 132$ and $x_K = 59$ is $x_K = 95.5$. Then $x_T = 153 - 95.5 - 18 = 39.5$. Hence $\nu = (39.5, 18, 95.5)$.

Finally, the Gately point divides the payoffs proportionally to $v(N) - v(N - i)$, which in this case gives $153 - 77 : 153 - 117 : 153 - 21$ or $76 : 36 : 132$ for the payoffs to T, A, K. Thus for example $x_T = [76/(76 + 36 + 132)] \cdot 153 = 47.7$. We get $\gamma = (47.7, 22.6, 82.8)$.

The Shapley value, the nucleolus and the Gately point are all plotted in Figure 32.1. We see that the nucleolus is most favorable to Kerala-Mysore, giving it the most credit for its strategic hydroelectric capabilities. The Shapley value splits the savings more equally, treats Andhra Pradesh generously, and lies near the Tamil Nadu/Kerala-Mysore constraint of the core. The Gately point is intermediate. If we translate these savings imputations back into costs by subtracting them from the go-it-alone costs $(533, 187, 86)$, the results are as follows:

	Cost allocated to		
Allocation method	T	A	K
Incurred cost	301	101	251
Proportional to power use	320	131	202
Shapley value	485	159	10
Gately point	485	164	3
Nucleolus	493	169	−9

Although the game theoretic cost allocations differ slightly, their import is clear. Tamil Nadu should pay about three-quarters of the joint project cost, Andhra Pradesh the other quarter, and Kerala-Mysore should essentially get a free ride. The hydroelectric resources of Kerala-Mysore are so important that they result in very significant savings to whatever coalition they are available to. The game theoretic solutions recognize and reward the resulting strength of Kerala-Mysore's bargaining position.

Game theoretic methods play an important role in modern cost allocation. [Straffin and Heaney, 1981] discusses allocation methods investigated by the Tennessee Valley Authority (TVA) in the 1930's, which turn out to be closely related to modern game theoretic ideas. The survey articles [Lucas, 1981] and [Young, 1985a] and the book [Young, 1985b] include thorough bibliographies covering both theory and applications. [Gately, 1974b] applies this kind of analysis to the East African Common Market formed by Kenya, Uganda, and Tanganyika in the 1950's. [Young, Okada, and Hashimoto, 1982] analyzes a cooperative water resources development project in Sweden as a six-person game.

33. The Value of Game Theory

When I ask my students who enjoy mathematics *why* they enjoy it, one of the most common answers I get is that mathematics "gives exact answers." It offers comforting certainty as opposed to the tenuousness and ambiguity of the humanities and the qualitative social sciences. It is, then, a natural hope that when mathematical reasoning is applied to the study of human interactions, as it is in game theory, exact answers will emerge. What was cloudy and ambiguous before will become clear and certain. Hope like this animates A. H. Copeland's famous review of von Neumann and Morgenstern's *Theory of Games and Economic Behavior*:

> Posterity may regard this book as one of the major scientific achievements of the first half of the twentieth century. This will undoubtedly be the case if the authors have succeeded in establishing a new exact science—the science of economics. The foundation which they have laid is extremely promising. [Copeland, 1945]

The first major result of our study of game theory was such an exact answer. What is rational behavior for two players in a pure conflict situation? Von Neumann's minimax theorem gives a convincing answer: they should play their optimal pure or mixed strategies, thereby obtaining the value of the game. Of course, the context here involves assumptions about utility, knowledge, and definiteness of strategies and outcomes. Within that context, though, the answer is pleasingly certain.

Unfortunately, as soon as we consider situations where conflict is against nature, or conflict is mixed with opportunities for cooperation, or more than two players are involved, this exact answer dissolves into a multiplicity of competing ideas which is distinctly uncomforting to anyone in search of certainty. Let us review the situation briefly.

We first met multiple solutions in Chapter 10, when a rational player plays a game against nature. The methods of Laplace, Hurwicz, Wald, and Savage offer different and conflicting advice about how to proceed. We can clarify the situation, as Milnor did, by axiomatic study of these methods, but we cannot point to one method and say it is unconditionally best.

In two-person non-zero-sum games the problems of multiple, non-interchangeable and non-equivalent equilibria destroy the uniqueness of the zero-sum prescription for play. The most serious problem, though, is the possibility of equilibria not being Pareto optimal. The purest form of the resulting dilemma is the Prisoner's Dilemma, with its absolute conflict between individual rationality and collective rationality. Once the dilemma is identified in pure form, we find versions of it in many areas of human interaction. The dilemma seems to be pervasive. We looked at several attempts to provide a logical escape from the dilemma, but none is completely convincing.

When we try to provide fair arbitration for non-zero-sum games, the Nash arbitration scheme offers an attractive first pass at a solution, but the problem of how the status quo point should be chosen produces different and competing answers.

N-person game theory seems to glory in multiple and conflicting approaches, even after we follow von Neumann and Morgenstern's advice and concentrate on the cooperative possibilities of coalitions via the characteristic function form. Von Neumann and Morgenstern's stable sets, the core, and Aumann and Maschler's bargaining sets provide complicated, overlapping, and conflicting prescriptions for rational outcomes. Worse, they offer differing approaches to how we should interpret bargaining in a rational context.

When we ask for a line of reasoning that leads to a single division of payoffs which would be fair, the Shapley value has a preeminent position. Yet even here the situation is not clearcut. In the context of measuring power in voting games, the Banzhaf index offers competition, and I have argued that there are cases in which it is more appropriate than the Shapley-Shubik power index (i.e. the Shapley value). In more general contexts the nucleolus, the Gately point and other related ideas are based more directly on various concepts of bargaining and threats. For cost allocation problems, these different lines of reasoning lead to different recommendations.

What should we make of this? In particular, should we consider it a failure of game theory to offer unique, clearcut prescriptions for analysis or action in these situations? Perhaps what began as an attempt to clarify the logic of conflict and cooperation has only deepened the confusion.

I would argue that game theory has been and is a useful and powerful tool for the understanding of human affairs. My argument is easier because I start with a different perspective from those who value mathematics because of its exactness and certainty. The pleasure I take in mathematics is much more in the process than in the result. Solving a problem or proving a theorem does indeed offer a thrill of satisfaction, but the most powerful reward is in the play of ideas, deadends, new approaches, breakthrough insights that eventually lead to the answer or the proof. Since I think that the ideas of game theory—the fascination of the Prisoner's Dilemma, the elegance of the Nash arbitration scheme, the subtlety of von Neumann-Morgenstern stable sets, the interplay between axioms and procedure in the Shapley value—are exciting in themselves, I am not discouraged that they fail to lead to exact and incontrovertible answers to the social problems they are designed to model.

Of course, my aesthetic feelings would be out of place—a mathematician's reveling in the ivory tower—if there *were* exact and incontrovertible answers to those problems, and game theory simply failed to find them. Surely, though, we know that human social interactions can be immensely complicated, rich in variety and meaning which cannot be captured by logic alone. What game theory reveals is that the underlying logical structure of conflict and cooperation is also complicated, rich and various. Individual self-interest and social welfare can be incompatible, and the Prisoner's Dilemma helps us understand why, with impressive clarity. In some situations coalitional bargaining is inherently unstable, and the concept of the core of a game and its possible emptiness sheds light on when and why that is true. The idea of power in political science, or fairness in welfare economics, is not logically simple, and the mathematical study of power indices or allocation methods clarifies the logical complexity. Game theory is complex, in other words, because it models, not unfaithfully, an underlying complex logical reality. Or so I would argue.

Anatol Rapoport, in a marvelously provocative essay at the end of *Mathematical Models in the Social and Behavioral Sciences*, takes this line of reasoning one step further. In the study of human affairs, he argues, the greatest enemy is dogma: the certainty that we have found the right, simple answer to a question which has no right, simple answer. Mathematical analysis of the kind embodied in game theory is the enemy of dogma:

> Mathematical thinking ... embodies a remarkable synthesis of maximum freedom and maximum discipline, something extremely difficult to combine in human affairs, where 'freedom' and 'discipline' almost always appear to be in opposition. [It] confers freedom from encrusted thinking habits, above all freedom from the 'tyranny of words,' which forces people to think in terms of culturally inherited or politically imposed verbal categories that need not be either internally consistent or related to anything in the real world ... Tyranny thrives on the canonization of nonsense and on the perpetuation of falsehood. Scientific thinking is a powerful antidote to tyranny ... It steers thinking away from slogans toward analysis. [Rapoport, 1983]

If this book has done its job well, you as reader should be healthily immune to calls to follow "the rational" course of action in situations of mixed conflict and cooperation, or settle on "the fair" distribution of benefits. You must ask for clarification, and you should have available to you a useful collection of concepts, models and lines of reasoning to support your skepticism.

I would add an emphasis that game theory offers more than just an antidote to simplistic error. It also contributes a wide range of clarifying insights, as I hope the variety of applications we have considered illustrates. The goal in the study of social interactions cannot be simple answers. I think it should be insight: new ways of analyzing, organizing and appreciating human experience. Game theory has been, and gives every indication of continuing to be, a powerful tool for the generation of insight.

Bibliography

Adorno, T. W., E. Frenkel-Brunswick, D. Levinson, and R. Sanford, *The Authoritarian Personality*, Harper, 1950.

Allen, Layman, "Games bargaining: a proposed application of the theory of games to collective bargaining," *Yale Law Journal* 65 (1956) 660–693.

Aumann, Robert, "A survey of games without sidepayments," pages 3–27 in M. Shubik, ed., *Essays in Mathematical Economics in Honor of Oskar Morgenstern*, Princeton University Press, 1967.

Aumann, Robert, and M. Maschler, "The bargaining set for cooperative games," pages 443–476 in Dresher, Shapley, and Tucker, eds., *Advances in Game Theory*, Princeton University Press, 1964.

Axelrod, Robert, *Conflict of Interest*, Markham Publishing Company, 1970.

Axelrod, Robert, *The Evolution of Cooperation*, Basic Books, 1984.

Axelrod, Robert, and W. D. Hamilton, "The evolution of cooperation," *Science* 212 (1981) 1390–1396.

Bacharach, Michael, *Economics and the Theory of Games*, Westview Press, 1977.

Bagnato, Robert, "A reinterpretation of Davenport's game theory analysis," *American Anthropologist* 76 (1974) 65–66.

Banzhaf, John, "Weighted voting doesn't work: a mathematical analysis," *Rutgers Law Review* 19 (1965) 317–343.

Bar-Hillel, M., and A. Margalit, "Newcomb's paradox revisited," *British Journal of the Philosophy of Science* 23 (1972) 295–304.

Barth, Fredrik, "Segmentary opposition and the theory of games: a study of Pathan organization," *Journal of the Royal Anthropological Institute* 89 (1959) 5–21.

Beresford, R. S., and M. H. Peston, "A mixed strategy in action," *Operational Research* 6 (1955) 173–175.

Billera, Louis, "Economic market games," pages 37–53 in W. Lucas, ed., *Game Theory and its Applications*, American Mathematical Society, 1981.

Braithwaite, R. B., *The Theory of Games as a Tool for the Moral Philosopher*, Cambridge University Press, 1955.

Brams, Steven, *Game Theory and Politics*, Free Press, 1975a.

Brams, Steven, "Newcomb's problem and prisoner's dilemma," *Journal of Conflict Resolution* 19 (1975b) 596–619.

Brams, Steven, *Paradoxes in Politics*, Free Press, 1976.

Brams, Steven, "The network television game: there may be no best schedule," *Interfaces* 7 (1977a) 102–109.

Brams, Steven, "Deception in 2 × 2 games," *Journal of Peace Science* 2 (1977b) 171–203.

Brams, Steven, *The Presidential Election Game*, Yale University Press, 1978.

Brams, Steven, *Biblical Games: A Strategic Analysis of Stories in the Old Testament*, MIT Press, 1980.

Brams, Steven, *Superpower Games*, Yale University Press, 1985.

Brams, Steven, *Negotiation Games*, Routledge, Chapman and Hall, 1990.

Brams, Steven, M. Davis, and P. Straffin, "The geometry of the arms race," *International Studies Quarterly* 23 (1979) 567–588.

Brams, Steven, and Peter Fishburn, *Approval Voting*, Birkhauser, 1983.

Brams, Steven, and M. Hessel, "Absorbing outcomes in 2 × 2 games," *Behavioral Science* 27 (1982) 393–401.

Brams, Steven, and D. M. Kilgour, *Game Theory and National Security*, Basil Blackwell, 1988.

Brams, Steven, and Philip Straffin, "Prisoner's dilemma and professional sports drafts," *American Mathematical Monthly* 86 (1979) 80–88.

Christie, R., and M. Jahoda, *Studies in the Scope and Method of 'The Authoritarian Personality,'* Free Press, 1954.

Colman, Andrew, *Game Theory and Experimental Games*, Pergamon Press, 1982.

Converse, P., and H. Valen, "Dimensions of change and perceived party distance in Norwegian voting," *Scandinavian Political Studies* 6 (1971) 107–151.

Copeland, A. H, review of von Neumann and Morgenstern's *Theory of Games and Economic Behavior*, *Bulletin of the American Mathematical Society* 51 (1945) 498–504.

Dr. Crypton, "Perils of the football draft," *Science Digest* (July 1986) 76–79.

Davenport, W. C., "Jamaican fishing: a game theory analysis," pages 3–11 in I. Rouse, ed., *Papers in Caribbean Anthropology #59*, Yale University Press, 1960.

Davis, Morton, *Game Theory: A Non-Technical Introduction*, Basic Books, 1983.

Dawkins, Richard, *The Selfish Gene*, Oxford University Press, 1976.

DeSwaan, Abraham, *Coalition Theories and Cabinet Formation*, Elsevier, 1973.

Deutsch, Morton, "Trust and suspicion," *Journal of Conflict Resolution* 2 (1958) 265–279.

Deutsch, Morton, "Trust, trustworthiness and the *F*-scale," *Journal of Abnormal and Social Psychology* 61 (1960) 138–140.

Dodd, L., *Coalitions in Parliamentary Government*, Princeton University Press, 1976.

Dubey, Pradeep, "On the uniqueness of the Shapley value," *International Journal of Game Theory* 4 (1975) 131–139.

Dubey, Pradeep, and Lloyd Shapley, "Mathematical properties of the Banzhaf index," *Mathematics of Operations Research* 4 (1979) 99–131.

Enelow, J., and M. Hinich, *The Spatial Theory of Voting: An Introduction*, Cambridge University Press, 1984.

Farquharson, Robin, *Theory of Voting*, Yale University Press, 1969.

Fisher, Roger, and William Ury, *Getting to Yes*, Houghton Mifflin, 1981.

Fouraker, L., and S. Siegel, *Bargaining and Group Decision Making*, McGraw-Hill, 1960.

Friedman, James, *Game Theory with Applications to Economics*, Oxford University Press, 1986.

Gardner, Martin, "Mathematical games," *Scientific American* (July 1973 and March 1974).

Gately, Dermot, "Sharing the gains from regional cooperation: a game theoretic application to planning investment in electric power," *International Economic Review* 15 (1974a) 195–208.

Gately, Dermot, "Sharing the gains from customs unions among less developed countries: a game theoretic approach," *Journal of Development Economics* 1 (1974b) 213–233.

Gibbard, A., "Manipulation of voting schemes: a general result," *Econometrica* 41 (1973) 587–602.

Grotte, J. H., "Observations on the nucleolus and the central game," *International Journal of Game Theory* 1 (1971/72) 173–177.

Hamburger, Henry, "*N*-person prisoner's dilemma," *Journal of Mathematical Sociology* 3 (1973) 27–48.

Hamburger, Henry, *Games as Models of Social Phenomena*, W. H. Freeman, 1979.

Hamlen, S., W. Hamlen, and J. Tschirhart, "The use of core theory in evaluating joint cost allocation schemes," *Accounting Review* 52 (1977) 616–627.

Hardin, Garrett, "The tragedy of the commons," *Science* 162 (1968) 1243–1248.

Harnett, D., "Bargaining and negotiation in a mixed motive game: price leadership in a bilateral monopoly," *Southern Economic Journal* 33 (1967) 479–487.

Haywood, O. G., "Military decision and game theory," *Journal of the Operations Research Society of America* 2 (1954) 365–385.

Heaney, James, "Urban wastewater management," Chapter 5 in Brams, Lucas, and Straffin, eds., *Political and Related Models*, Springer-Verlag, 1983.

Herstein, I. N., and John Milnor, "An axiomatic approach to measurable utility," *Econometrica* 21 (1953) 291–297.

Hill, W. W., "Prisoner's dilemma, a stochastic solution," *Mathematics Magazine* 48 (1975) 103–105.

Hofstadter, Douglas, "Metamagical themas," *Scientific American* 248 (May 1983) 16–26.

Holler, M., ed., *Coalition Theory*, Physica-Verlag, 1983.

Howard, Nigel, *Paradoxes of Rationality: Theory of Metagames and Political Behavior*, MIT Press, 1971.

Johnson, S. M., "A game solution to a missile penetration problem," pages 250–267 in A. Mensch, ed., *The Theory of Games: Techniques and Applications*, Elsevier, 1966.

Jones, A. J., *Game Theory: Mathematical Models of Conflict*, Halsted Press, 1980.

Kalai, E., and M. Smorodinsky, "Other solutions to Nash's bargaining problem," *Econometrica* 43 (1975) 513–518.

Kilgour, D. M., "A formal analysis of the amending formula of Canada's Constitution Act, 1982," *Canadian Journal of Political Science* 16 (1983) 771–777.

Kilgour, D. M., and T. J. Levesque, "The Canadian constitutional amending formula: bargaining in the past and the future," *Public Choice* 44 (1984) 457–480.

Kohler, D., and R. Chandrasekaran, "A class of sequential games," *Operations Research* 19 (1971) 270–277.

Kozelka, Robert, "A Bayesian approach to Jamaican fishing," in Buchler and Nutini, eds., *Game Theory in the Behavioral Sciences*, University of Pittsburgh Press, 1969.

Kreps, David, *A Course in Microeconomic Theory*, Princeton University Press, 1990a.

Kreps, David, *Game Theory and Economic Modelling*, Oxford University Press, 1990b.

Lambert, John, "Voting games, power indices, and presidential elections," pages 144–197 in Paul Campbell, ed., *UMAP Modules: Tools for Teaching, 1988*, Consortium for Mathematics and Its Applications, 1989.

Levi, Isaac, "Newcomb's many problems," *Theory and Decision* 6 (1975) 161–175.

Littlechild, S., and K. Vaidya, "The propensity to disrupt and the disruption nucleolus of a characteristic function game," *International Journal of Game Theory* 5 (1976) 151–161.

Lucas, William, "A game with no solution," *Bulletin of the American Mathematical Society* 74 (1968) 237–239.

Lucas, William, "Multiperson cooperative games," pages 1–17 in W. F. Lucas, ed., *Game Theory and Its Applications*, American Mathematical Society, 1981.

Lucas, William, "Applications of cooperative games to equitable allocation," pages 19–36 in W. Lucas, ed., *Game Theory and Its Applications*, American Mathematical Society, 1981.

Lucas, William, "Measuring power in weighted voting systems," Chapter 9 in Brams, Lucas, and Straffin, eds., *Political and Related Models*, Springer-Verlag, 1983.

Lucas, William, and L. J. Billera, "Modeling coalitional values," Chapter 4 in Brams, Lucas, and Straffin, eds., *Political and Related Models*, Springer-Verlag, 1983.

Luce, R. D., and Howard Raiffa, *Games and Decisions*, Wiley, 1957.

Maurer, Stephen, "An interview with Albert W. Tucker" and "The mathematics of Tucker: a sampler," *The Two-Year College Mathematics Journal* 14 (1983) 210–224, 228–232.

Mayberry, J., J. Nash, and M. Shubik, "Comparison of treatments of a duopoly situation," *Econometrica* 21 (1953) 141–154.

Maynard Smith, John, "Evolution and the theory of games," *American Scientist* 64 (1976) 41–45.

Maynard Smith, John, "Game theory and the evolution of behavior," *Proceedings of the Royal Society London* B205 (1979) 475–488.

Maynard Smith, John, *Evolution and the Theory of Games*, Cambridge University Press, 1982.

Maynard Smith, John, and G. R. Price, "The logic of animal conflict," *Nature* 246 (1973) 15–18.

McDonald, John, *The Game of Business*, Anchor Books, 1977.

McDonald, John, and John Tukey, "Colonel Blotto: a problem in military strategy," *Fortune* (June 1949) 102. Reprinted on pages 226–229 in M. Shubik, ed., *Game Theory and Related Approaches to Social Behavior*, Wiley, 1964.

McKelvey, R., P. Ordeshook, and M. Winer, "The competitive solution for *n*-person games without transferable utility, with an application to committee games," *American Political Science Review* 72 (1978) 599–615.

McKinsey, J. C., *Introduction to the Theory of Games*, RAND Corporation, 1952.

Miller, D., "A Shapley value analysis of the proposed Canadian constitutional amendment scheme," *Canadian Journal of Political Science* 4 (1973) 140–143.

Milnor, John, "Games against nature," pages 49–59 in Thrall et al., eds., *Decision Processes*, Wiley, 1954. Reprinted on pages 120–131 in M. Shubik, ed., *Game Theory and Related Approaches to Social Behavior*, Wiley, 1964.

Moore, O. K., "Divination: a new perspective," *American Anthropologist* 59 (1957) 69–74.

Myerson, R., *Game Theory: Analysis of Conflict*, Harvard University Press, 1991.

Nash, John, "Equilibrium points in *n*-person games," *Proceedings of the National Academy of Sciences USA* 36 (1950a) 48–49.

Nash, John, "The bargaining problem," *Econometrica* 18 (1950b) 155–162.

Nash, John, "Non-cooperative games," *Annals of Mathematics* 54 (1951) 286–299.

Nash, John, "Two-person cooperative games," *Econometrica* 21 (1953) 128–140.

New York Times, "'Democrats' Contra aid plan defeated by House, 216–208; all U.S. funds now halted," March 4, 1988.

Nozick, Robert, "Newcomb's problem and two principles of choice," pages 114–146 in N. Rescher et al., eds., *Essays in Honor of Carl G. Hempel*, D. Reidel, 1969.

Owen, Guillermo, "Political games," *Naval Research Logistics Quarterly* 18 (1971) 345–354.

Owen, Guillermo, "Evaluation of a presidential election game," *American Political Science Review* 69 (1975) 947–953.

Owen, Guillermo, *Game Theory* (2nd edition), Academic Press, 1982.

Parkinson, Pauline, "The power of voting blocs," term paper for Game Theory course, Beloit College, 1981.

Polgreen, Philip, "An application of Nash's arbitration scheme to a labor dispute," *UMAP Journal* 13 (1992) 25–35.

Ponssard, Jean-Pierre, *Competitive Strategies*, North-Holland, 1981.

Poundstone, William, *Prisoner's Dilemma: John von Neumann, Game Theory, and the Puzzle of the Bomb*, Doubleday, 1992.

Rabinowitz, G., and S. Macdonald, "The power of the states in U.S. presidential elections," *American Political Science Review* 80 (1986) 65–87.

Raiffa, Howard, *The Art and Science of Negotiation*, Harvard University Press, 1982.

Rapoport, Anatol, "Exploiter, leader, hero and martyr: the four archetypes of the 2×2 game," *Behavioral Science* 12 (1967a) 81–84.

Rapoport, Anatol, "Escape from paradox," *Scientific American* 217 (1967b) 50–56.

Rapoport, Anatol, *Two-Person Game Theory: The Essential Ideas*, University of Michigan Press, 1970a.

Rapoport, Anatol, *N-Person Game Theory: Concepts and Applications*, University of Michigan Press, 1970b.

Rapoport, Anatol, "Prisoner's Dilemma—recollections and observations," pages 17–34 in A. Rapoport, ed., *Game Theory as a Theory of Conflict Resolution*, D. Reidel, 1974.

Rapoport, Anatol, *Mathematical Models in the Social and Behavioral Sciences*, Wiley-Interscience, 1983.

Rapoport, Anatol, and Albert Chammah, "The game of chicken," pages 151–175 in I. Buchler and H. Nutini, eds., *Game Theory in the Behavioral Sciences*, University of Pittsburgh Press, 1969.

Rapoport, Anatol, and Albert Chammah, *Prisoner's Dilemma*, University of Michigan Press, 1970.

Rapoport, Anatol, and M. Guyer, "A taxonomy of 2×2 games," *General Systems* 11 (1966) 203–214.

Rapoport, Anatol, M. Guyer, and D. Gordon, *The* 2×2 *Game*, University of Michigan Press, 1976.

Rasmusen, Eric, *Games and Information: An Introduction to Game Theory*, Basil Blackwell, 1989.

Read, D. W., and C. E. Read, "A critique of Davenport's game theory analysis," *American Anthropologist* 72 (1970) 351–355.

Riker, William, "The paradox of voting and Congressional rules for voting on amendments," *American Political Science Review* 52 (1958) 349–366.

Riker, William, *The Theory of Political Coalitions*, Yale University Press, 1962.

Riker, William, "Arrow's theorem and some examples of the paradox of voting," in J. Claunch, ed., *Mathematical Applications in Political Science*, Arnold Foundation of SMU, 1965.

Riker, William, and Peter Ordeshook, *An Introduction to Positive Political Theory*, Prentice Hall, 1973.

Roberts, Fred S., *Measurement Theory*, Addison-Wesley, 1979.

Robinson, Julia, "An iterative method of solving a game," *Annals of Mathematics* 54 (1951) 296–301.

Roth, Alvin, ed., *The Shapley Value: Essays in Honor of Lloyd S. Shapley*, Cambridge University Press, 1988.

Satterthwaite, M., "Strategyproofness and Arrow's conditions: existence and correspondence theorems for voting procedures and social welfare functions," *Journal of Economic Theory* 10 (1975) 187–217.

Savage, L. J., *The Foundations of Statistics*, Wiley, 1954. Second revised edition, Dover, 1972.

Schelling, Thomas, *The Strategy of Conflict*, Oxford University Press, 1960.

Schelling, Thomas, *Micromotives and Macrobehavior*, Norton, 1978.

Schmeidler, David, "The nucleolus of a characteristic function game," *SIAM Journal on Applied Mathematics* 17 (1969) 1163–1170.

Schofield, Norman, "The kernel and payoffs in European government coalitions," *Public Choice* 26 (1976) 29–49.

Shapley, L. S., "A value for n-person games," pages 307–317 in Kuhn and Tucker, eds., *Contributions to the Theory of Games, II*, Princeton University Press, 1953.

Shapley, Lloyd, "Simple games: an outline of the descriptive theory," *Behavioral Science* 7 (1962) 59–66.

Shapley, Lloyd, "A comparison of power indices and a non-symmetric generalization," RAND Paper P-5872, Santa Monica, 1977.

Shapley, Lloyd, "Valuation of games," pages 55–67 in W. Lucas, ed., *Game Theory and its Applications*, American Mathematical Society, 1981.

Shapley, Lloyd, "Measurement of power in political systems," pages 69–81 in W. Lucas, ed., *Game Theory and its Applications*, American Mathematical Society, 1981.

Shapley, Lloyd, and Martin Shubik, "A method for evaluating the distribution of power in a committee system," *American Political Science Review* 48 (1954) 787–792.

Shapley, Lloyd, and M. Shubik, "On the core of an economic system with externalities," *American Economic Review* 59 (1969) 678–684.

Shubik, Martin, "Incentives, decentralized control, the assignment of joint costs and internal pricing," *Management Science* 8 (1962) 325–343.

Shubik, Martin, "Game theory, behavior, and the paradox of prisoner's dilemma: three solutions," *Journal of Conflict Resolution* 14 (1970) 181–194.

Shubik, Martin, *Game Theory and the Social Sciences: Concepts and Solutions*, MIT Press, 1982.

Siegel, S., and D. Harnett, "Bargaining behavior: a comparison between mature industrial personnel and college students," *Operations Research* 12 (1964) 334–343.

Straffin, Philip, "The power of voting blocs: an example," *Mathematics Magazine* 50 (1977a) 22–24.

Straffin, Philip, "The bandwagon curve," *American Journal of Political Science* 21 (1977b) 695–709.

Straffin, Philip, "Homogeneity, independence and power indices," *Public Choice* 30 (1977c) 107–118.

Straffin, Philip, *Topics in the Theory of Voting*, Birkhauser, 1980a.

Straffin, Philip, "The prisoner's dilemma," *UMAP Journal* 1 (1980b) 101–103.

Straffin, Philip, "Power indices in politics," Chapter 11 in Brams, Lucas, and Straffin, eds., *Political and Related Models*, Springer-Verlag, 1983.

Straffin, Philip, "Three person winner-take-all games with McCarthy's revenge rule," *College Mathematics Journal* 16 (1985) 386–394.

Straffin, Philip, "The Shapley-Shubik and Banzhaf indices as probabilities," in A. Roth, ed., *The Shapley Value: Essays in Honor of Lloyd S. Shapley*, Cambridge University Press, 1988.

Straffin, Philip, "Game theory and nuclear deterrence," *UMAP Journal* 10 (1989a) 87–92.

Straffin, Philip, "Spatial models of power and voting outcomes," pages 315–335 in Fred Roberts, ed., *Applications of Combinatorics and Graph Theory in the Biological and Social Sciences*, Springer-Verlag, 1989b.

Straffin, Philip, and Bernard Grofman, "Parliamentary coalitions: a tour of models," *Mathematics Magazine* 57 (1984) 259–274.

Straffin, Philip, and James Heaney, "Game theory and the Tennessee Valley Authority," *International Journal of Game Theory* 10 (1981) 35–43.

Thomas, L. C., *Games, Theory and Applications*, Halsted Press, 1984.

Von Neumann, John, and Oskar Morgenstern, *Theory of Games and Economic Behavior*, Wiley, 1967 (original edition 1944).

Vorob'ev, N. N., *Game Theory*, Springer-Verlag, 1977.

Wald, Abraham, *Statistical Decision Functions*, Wiley, 1950.

Weber, Robert, "Non-cooperative games," pages 83–125 in W. Lucas, ed., *Game Theory and Its Applications*, American Mathematical Society, 1981.

Williams, J. D., *The Compleat Strategyst*, Dover, 1986. (Original edition by RAND Corporation, 1954)

Young, H. P., "Cost allocation," pages 65–94 in H. P. Young, ed., *Fair Allocation*, American Mathematical Society, 1985a.

Young, H. P., ed., *Cost Allocation: Methods, Principles, Applications*, North Holland, 1985b.

Young, H. P., N. Okada, and T. Hashimoto, "Cost allocation in water resources development," *Water Resources Research* 18 (1982) 463–475.

Zermelo, Ernst, "Uber eine Anwendung der Mengenlehre und der Theorie des Schachspiels," pages 501–504 in *Proceedings of the Fifth International Congress of Mathematicians*, Cambridge, 1912.

Answers to Exercises

2.1) Colin B dominates Colin D; Colin C dominates Colin A.

2.2) Colin C dominates Colin A and B, so cross out Colin A and B. Then Rose A dominates Rose B and C, so cross out Rose B and C. Then Colin E dominates Colin D, so cross out Colin D. The resulting reduced game involves Rose A and D, Colin C and E.

2.3) a) Four saddle points at Rose A or C, Colin B or D.

b) Saddle point at Rose B–Colin B.

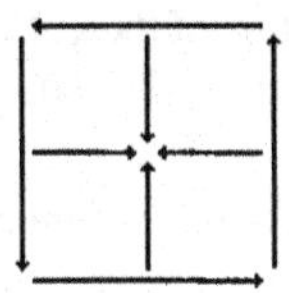

c) No saddle point.

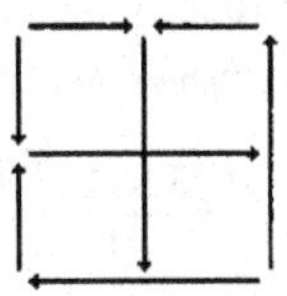

2.4) Since b is largest in its column, $b \geqslant a$. Since Rose A dominates Rose B, $a \geqslant b$. Hence $a = b$, and a is also a largest entry in its column. To show that a is a smallest entry in its row, consider any other entry c in Rose A and let d be the corresponding entry in Rose B. Then $c \geqslant d \geqslant b = a$, where the first inequality holds by dominance and the second holds because b is a saddle point.

2.5) a) Since a is smallest in its row, it is the row minimum for its row. Since it is largest in its column, the other row minima cannot be larger. Hence it is the row maximin. Similarly, it is the column minimax.

b) Let a be the entry in the maximin row I and the minimax column J. Then the minimum of Rose I $\leqslant a \leqslant$ the maximum of Colin J. Since we are given that the two extreme numbers in the inequality are the same, the inequality is in fact an equality. Hence a is smallest in its row and largest in its column.

2.6) a) If I were Rose, I would calculate the *expected value* of playing each of my strategies, and play the one for which the expected value is largest. The expected value is a weighted average of possible payoffs, where the weights are the probabilities that each outcome will occur. For instance, the expected value of Rose A is $.20(12) + .51(-1) + .02(1) + .27(0) = 1.91$. The expected values for Rose B, C, D are -3.75, 2.51, 1.12. I would play Rose C. The expected value will be an important idea in Chapter 3.

b) I would play Colin B, whose expected value of .77 is smaller than the other expected values of 4.09, 2.97, 1.07. Remember that Colin wants the payoff to be small.

2.7) No. See the movement diagram for the game in Exercise 2.3b.

3.2) a) x may not lie between 0 and 1.

b) The oddments will not yield a solution.

3.3) a) Rose $(0, \frac{1}{2}, \frac{1}{2}, 0)$; Colin $(\frac{5}{8}, \frac{3}{8})$; value $= \frac{1}{2}$.

b) Saddle point at Rose B, Colin A; value = 1.

c) Rose $(\frac{4}{9}, \frac{5}{9})$; Colin $(0, 0, 0, \frac{1}{3}, \frac{2}{3})$; value $= -\frac{1}{3}$.

3.4) a) The three lines intersect at one point.

b) The value is $\frac{1}{2}$, and Rose's optimal strategy is $(\frac{1}{2}, \frac{1}{2})$. Colin can play either $(\frac{3}{8}, 0, \frac{5}{8})$ or $(0, \frac{3}{4}, \frac{1}{4})$, or any mixture of these. Colin A and B doesn't yield a solution because both of those lines slant to the left; the solution to that 2×2 subgame would be a saddle point.

3.5) a) Rose C is dominated; then Colin C is dominated. Rose $(\frac{1}{2}, \frac{1}{2}, 0)$; Colin $(\frac{1}{3}, \frac{2}{3}, 0)$; value = 1.

b) Saddle point at Rose C, Colin C; value = 3.

c) Eliminate first Colin A, then Rose A, then Colin D, and solve resulting 4×2 game. Rose $(0, 0, \frac{1}{2}, 0, \frac{1}{2})$; Colin $(0, \frac{5}{8}, \frac{3}{8}, 0)$; value $= -\frac{1}{2}$.

3.7) a) We use the notation S1G1 for "show one, guess one," etc.

	S1G1	S2G1	S1G2	S2G2
S1G1	0	−3	2	0
S2G1	3	0	0	−4
S1G2	−2	0	0	3
S2G2	0	4	−3	0

3.8) $(\frac{2}{7}, \frac{2}{7}, \frac{2}{7}, \frac{1}{7})$ is optimal for both players. The value is $\frac{12}{7}$.

4.1) If the current runs 30% of the time, In-Out has the highest expected value (13.46). To find such a percentage, look at Figure 4.1 and see where the In-Out line is uppermost.

4.2) The solution is now to fish Inside 81% of the time, Outside 19%. The In-Out strategy is inactive. The value of the game falls to 13.2, a very minor drop. If you haven't done so already, you should look at what effect the change had on the payoff diagram for the game.

5.1) Guerrillas use 6-0 and 4-2, each with probability $\frac{1}{2}$. Police use 6-1 and 4-3, each with probability $\frac{1}{2}$. Value $= \frac{3}{4}$.

5.2) Red uses WDDD and DDDW, each with probability $\frac{1}{2}$. Blue uses 1 and 3, each with probability $\frac{1}{2}$. Value $= \frac{1}{2}$.

5.3) b) Blotto 3100, 2200, 1111 with probabilities $(\frac{4}{5}, 0, \frac{1}{5})$. Enemy 3000, 2100, 1110 with probabilities $(\frac{2}{5}, \frac{2}{5}, \frac{1}{5})$. The value is $\frac{3}{5}$. The 0 oddment for Blotto's strategy 2200 is what looks strange, but solutions like this can happen. There are other solutions to this game as well. The solution given in [McDonald and Tukey, 1949] is not correct.

5.4) a) Guerrillas use 200, 110 with probabilities $(\frac{2}{3}, \frac{1}{3})$. Police use 220, 211 with probabilities $(\frac{2}{3}, \frac{1}{3})$. Value $= \frac{4}{9}$.

b)

		Police				
		400	310	220	211	Guerrilla optimal
	300	$\frac{2}{3}$	$\frac{2}{3}$	1	1	$\frac{1}{3}$
Guerrillas	210	1	$\frac{5}{6}$	$\frac{2}{3}$	$\frac{2}{3}$	$\frac{2}{3}$
	111	1	1	1	0	
Police optimal			$\frac{2}{3}$		$\frac{1}{3}$	Value $= \frac{7}{9}$

6.1) b) Choosing both is still the maximin strategy. Or if you really believe that prediction is impossible, so the Being could only be correct with probability .5, the expected value of "take both" exceeds the expected value of "take only Box #2."

6.2) a) The Expected Value Principle says you should take only Box #2 if you believe that the probability of the Being predicting correctly is larger than .5005, exactly as in the original formulation.

b) It would seem more reasonable to me to take the states of nature to be "Box #2 is empty" and "Box #2 has $1,000,000 in it." We are then back in the original situation, with the Dominance Principle applying and telling us to take both boxes.

7.2)

		Colin bets on		
		A, K, Q	A, K	A
	A, K, Q	0	$-\frac{3}{9}$	0
Rose calls on	A, K	$\frac{3}{9}$	$\frac{1}{9}$	$\frac{2}{9}$
	A	0	$\frac{2}{9}$	$\frac{4}{9}$

Rose's first and Colin's last strategies are dominated. The solution calls for Colin to always bet with an ace or king, and to bet $\frac{1}{4}$ of the time with a queen (bluffing). Rose should always call with an ace, and call $\frac{1}{2}$ of the time with a king. The value of the game is $\frac{1}{6}$ to Rose.

7.3) Khrushchev would always acquiesce; knowing this, Kennedy would blockade; knowing this, Khrushchev should not place missiles. In 1963 Khrushchev did place missiles, Kennedy blockaded, Khrushchev acquiesced. The players were uncertain about each others payoffs, and Khrushchev perhaps counted on this uncertainty—if Kennedy was not certain Khrushchev would prefer to acquiesce, he might not have been willing to risk the outcomes x or z. There is also the possibility, of course, that the given tree is too crude a model of the real situation, or that our preference orderings are incorrect.

7.4) Colin: A/JM, A/KM, A/LM, B/NP, B/OP, C/QS, C/QT, C/QU, C/QV, C/RS, C/RT, C/RU, C/RV. In particular, note that although there are 13 final branches in the tree, this is *not* why Colin has 13 strategies. It is just a coincidence.

7.5) a) RLL: -1.

d)

		Colin							
		L/LL	L/LR	L/RL	L/RR	R/LL	R/LR	R/RL	R/RR
	LL	-1	-1	2	2	-1	-1	3	3
Rose	LR	-1	-1	2	2	4	-2	4	-2
	RL	1	0	1	0	-1	-1	3	3
	RR	1	0	1	0	4	-2	4	-2

7.6) a) On the third line of the tree, Colin's first and second nodes are now in a single information set, as are his third and fourth nodes.

b) Since Colin's second choice now cannot be contingent on Rose's choice, only the 1st, 4th, 5th, and 8th strategies above are possible.

d) The 4×4 game has several solutions, one of which is Rose $(0, \frac{1}{4}, \frac{1}{2}, \frac{1}{4})$, Colin $(\frac{1}{2}, \frac{1}{2}, 0, 0)$. Since the value is now $\frac{1}{2}$ instead of -1, Colin's loss of information has cost him $\frac{3}{2}$.

7.7) There is now one additional possibility: that both players have a strategy which will force at least a draw, regardless of what the other player does. It is believed by most experts that this is the situation in chess.

8.1) a) Zeus' nodes are in separate information sets; all of Athena's nodes are in one information set.

b)

		Athena		
		Lo	Hi	
	Lo/Lo	23	18	
Zeus	Lo/Hi	16	16	
(small/large)	Hi/Lo	25	22	←saddle point
	Hi/Hi	18	20	

c) Athena would play its optimal mixed strategy in Game 8.1. Zeus would play Hi/Lo and get $22\frac{6}{7}$.

8.2) b)

		Athena		
		Lo	Hi	
	Lo/Lo	23	18	
Zeus	Lo/Hi	23	20	←saddle point
(Athena Lo/Athena Hi)	Hi/Lo	18	18	
	Hi/Hi	18	20	

c) Since the value of the game is 20 instead of 18, Zeus would be $2 million better off.

8.3) a) Zeus' nodes are in one information set; each of Athena's nodes is in its own information set.

b) The matrix game would be 2×16.

c) Athena will make the best choice at each of its four nodes. If Zeus chooses Lo, its expectation is $(\frac{1}{2})(8 + 28) = 18$; for Hi its expectation is $(\frac{1}{2})(16 + 16) = 16$. Hence Zeus should choose Lo.

d) Zeus would think it was playing Game 8.2, in which it can play either Lo or Hi. If it chose Hi, its expectation would decrease by $2 million. If I were Zeus, I'd choose Lo just on the chance of an Athena survey.

9.1) a) x b) indifferent c) $\frac{1}{2}v, \frac{1}{2}u$. d) indifferent—both have utility $\frac{300}{7}$.

9.2) a) $\frac{3}{7}$A, $\frac{4}{7}$B

b) The lotteries are as in 9.1d.

9.3) Many utility scales are possible. One gives the game

		Colin	
		A	B
Rose	A	2	4
	B	5	0

Rose optimal is $\frac{5}{7}, \frac{2}{7}$. Colin optimal is $\frac{4}{7}, \frac{3}{7}$. The value of the game to Rose is $\frac{20}{7}$ on this scale, which is less than the utility of s.

9.4) a) Zero-sum. E.g. transform Colin's utilities by $f(x) = \frac{1}{20}(x - 10)$.

b) Not zero-sum. When graphed, the points don't lie on a line.

9.5) a) No—fallacy #4.

b) No—fallacy #4.

c) Yes, ratios of differences are meaningful on a cardinal scale.

d) No—fallacy #3.

e) Yes, since both players prefer s to p.

10.1) Laplace says hold steady, Wald says diversify, Hurwicz with $\alpha > \frac{1}{2}$ says expand greatly, Savage says expand slightly.

10.2) a) Wald would then choose A.

b) Adding a large bonus to any column except Column B will change the Hurwicz recommendation.

10.3) The new regret matrix is

	A	B	C	D	Row maximum	
A	0	2	1	2	2	←smallest
B	1	3	0	2	3	
C	2	0	1	3	3	
D	1	1	1	3	3	
E	2	4	1	0	4	

Note that Savage judges the new strategy to be worse than any of the old ones, but its existence still alters which strategy is thought best.

10.4) a) Add large bonuses to all columns which do not contain the largest entry.

b) Duplicate, many times, the column containing the largest entry.

11.1) There is a mixed strategy Nash equilibrium where both Rose and Colin play ($\frac{1}{2}$A, $\frac{1}{2}$B), and both players get an expected payoff of 2.5. This is not Pareto optimal, since BA would be better for both players.

11.2) a) The game is not SSS, since AB and BA are two non-equivalent, non-interchangeable equilibria.

b) Not SSS, since the only equilibrium, AA, is not Pareto optimal.

c) It is SSS, with solution BC = $(3, 3)$.

d) Not SSS, since $(3, 3)$ now lies in the interior of the payoff polygon, and so is no longer Pareto optimal.

11.3) a) The given strategy assures that Rose will get $\frac{8}{7}$ regardless of what Colin does.

b) Colin's game has a saddle point at BB, so Colin's prudential strategy is Colin B, with security level of 1.

c) $\frac{8}{7}$ to Rose, $\frac{20}{7}$ to Colin, not Pareto optimal.

d) Rose counter-prudential is Rose B. Colin counter-prudential is Colin B.

Rose strategy	Colin strategy	Rose payoff	Colin payoff
prudential	prudential	$\frac{8}{7}$	$\frac{20}{7}$
prudential	counter-prud	$\frac{8}{7}$	$\frac{20}{7}$
counter-prud	prudential	2	1
counter-prud	counter-prud	2	1

11.4) a) Solution is AC = $(2, 4)$. BB = $(4, 2)$ is also an equilibrium, but it is not Pareto optimal.

b) Colin A is dominated by Colin B. Rose C is dominated by Rose B. With these crossed out, the equilibrium at BB becomes Pareto optimal, and would seem to have equal footing with AC.

c) I probably would recommend AC, but I could understand Rose feeling unhappy with it!

12.1) a) A/AA: choose A in the first game, and A in the second game regardless of what Colin did in the first game.

A/AB: choose A in the first game, and in the second game choose whatever Colin chose in the first game. Etc.

b)

	Colin							
	A/AA	A/AB	A/BA	A/BB	B/AA	B/AB	B/BA	B/BB
A/AA	0, 0	0, 0	−2, 1	−2, 1	−2, 1	−2, 1	−4, 2	−4, 2
A/AB	0, 0	0, 0	−2, 1	−2, 1	−1, −1	−1, −1	−3, 0	−3, 0
A/BA	1, −2	1, −2	−1, −1	−1, −1	−2, 1	−2, 1	−4, 2	−4, 2
A/BB	1, −2	1, −2	−1, −1	−1, −1	−1, −1	−1, −1	−3, 0	−3, 0
B/AA	1, −2	−1, −1	1, −2	−1, −1	−1, −1	−3, 0	−1, −1	−3, 0
B/AB	1, −2	−1, −1	1, −2	−1, −1	0, −3	−2, −2	0, −3	−2, −2
B/BA	2, −4	0, −3	2, −4	0, −3	−1, −1	−3, 0	−1, −1	−3, 0
B/BB	2, −4	0, −3	2, −4	0, −3	0, −3	−2, −2	0, −3	−2, −2

c)

		Colin	
		A/BB	B/BB
Rose	A/BB	$(-1, -1)$	$(-3, 0)$
	B/BB	$(0, -3)$	$(-2, -2)$

12.2) a) $(1-p)(1+p+p^2+\cdots+p^{m-1}) = 1-p^m$.

b) As m goes to infinity, p^m goes to zero.

12.3) Rose IV-Colin II is cooperative. Rose XVI-Colin IV is non-cooperative.

12.4) a) AB and BA are both Pareto optimal equilibria.

b) Rose $(\frac{6}{7}, \frac{1}{7})$; Colin $(\frac{6}{7}, \frac{1}{7})$. The payoff is $-\frac{2}{7}$ to each player.

12.5) Let A represent going to the artfair, B going to the ballgame.

		Colin	
		A	B
Rose	A	(1,2)	(−2,−2)
	B	(−1,−1)	(2,1)

or any game ordinally equivalent to this. For a discussion, see [Rapoport, 1967a].

12.6) A good source of ideas is [Hamburger, 1979].

14.1) a) Rose threatens "If A then B" or lowers AA from 3 to 1.

b) Nothing works for Rose in this game.

c) Rose threatens "If A then B" and promises "If B then A," or lowers AA from 2 to 0 and BB from 4 to 2.

d) Rose moves first, or commits to B by lowering AA from 2 to 0 and AB from 4 to 2. Alternatively Rose promises "If B then B," lowering only AB.

e) Rose makes Colin move first.

14.2) One of many possibilities is

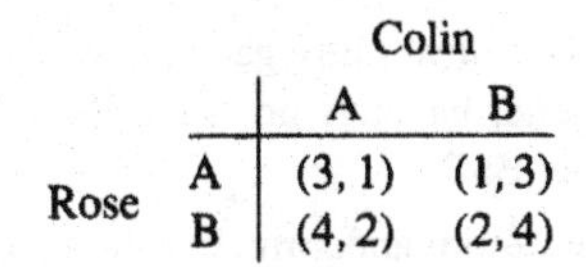

		Colin	
		A	B
Rose	A	(3, 1)	(1,3)
	B	(4,2)	(2,4)

14.3) a)

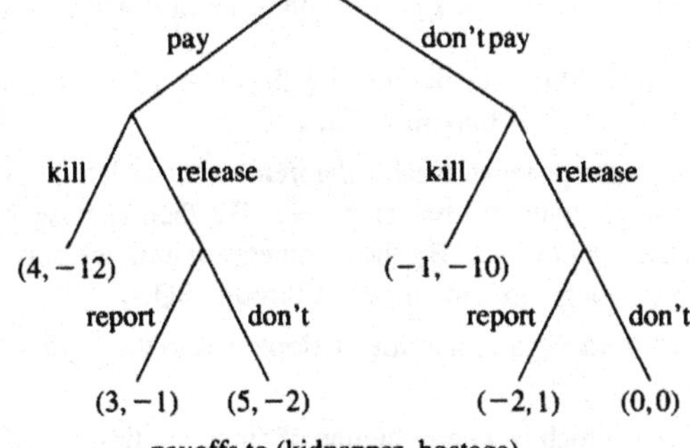

b) $(-1, -10)$, in which the hostage doesn't pay and is killed.

c) "If you release me, I will not report the kidnapping." Outcome is $(0, 0)$ and both benefit.

d) "If you don't pay, I will kill you." $(5, -2)$

e) "If you pay, I will release you." $(3, -1)$

f) Here are some suggestions from my students. For the hostage's promise: give the kidnapper incriminating information about something the hostage did. For the kidnapper's threat: act irrational and let slip that he has killed before. For the kidnapper's promise: "I'm doing this for someone else and I'll be in trouble if I don't get the money, but I don't want a murder on my hands."

14.4) a) $4! \cdot 4! = 576$

b) There are $2 \cdot 2$ ways to arrange the 4's, given that they must be in the A row or column. Then there are $3! \cdot 3!$ ways to fill in the other numbers.

c) There are six symmetric games with AA $= (4,4)$, and two each with AA $= (3,3), (2,2)$, or $(1,1)$.

15.2) B is the unique ESS in five games, A in one game, which is Prisoner's Dilemma. Three games have both A and B as ESS's, and three have only a mixed strategy ESS. Chicken is Game 15.4. (In other words, the hawk-dove game is Chicken...)

15.4) Hawk: $-25 + 75x$. Bully: $50x$. Dove or retaliator: 15. $50x > 15$ when $x > \frac{3}{10}$. Hawks always do worse than bullies.

15.5)

	hawk	dove	bully	retaliator
hawk	25	150	150	25
dove	0	65	0	65
bully	0	150	75	0
retaliator	25	65	150	65

In the hawk-dove or hawk-dove-bully game, hawk is the only ESS. In the hawk-dove-bully-retaliator game, hawk is not an ESS, since it would be invaded by retaliator. Retaliator is an ESS.

15.6) For the males, $\frac{5}{8}$ faithful, $\frac{3}{8}$ philandering. For the females, $\frac{5}{6}$ coy, $\frac{1}{6}$ fast. Expected payoffs are $1\frac{1}{4}$ for females, $2\frac{1}{2}$ for males. This is not Pareto optimal, since both sexes would do better in a population of faithful males and fast females.

16.1) b) The negotiation set lies on the line $y = 10 - \frac{1}{2}x$. Then $(x - \frac{10}{3})(y - 6) = -\frac{1}{2}x^2 + \frac{17}{3}x - \frac{40}{3}$ is maximized at $x = \frac{17}{3}$.

16.2) a) If, as in the proof of Nash's theorem, we translate SQ to $(0,0)$, the negotiation set of our polygon still has slope -1. We then enclose it in a polygon which is symmetric about $y = x$. By the symmetry axiom, the solution must lie on the line $y = x$, i.e. along a line of slope $+1$ through SQ.

b) Move from SQ along a line of slope $+m$ to the negotiation set. To see this, use Axiom 2.

c) $(4.5, 4)$, which is where the line of slope $+2$ through SQ $= (3,1)$ intersects the line $y = 13 - 2x$.

16.3) CD is in the line $y = 14 - x$. $(x - 2)(14 - x - 1)$ has its maximum at $x = 7.5$, which is to the left of C.

16.4) a) $D = (16, 10)$

b) $(4\frac{1}{2}, 6) = \frac{3}{4}B + \frac{1}{4}A$.

16.5) a) Security level SQ is $(2, 1)$, with Nash point $(5, 4)$. Threat strategy SQ is $(0, 1)$, with Nash point $(4, 5)$. Rose has a higher security level, but Colin has a stronger threat.

b) Security level SQ is $(0, 0)$, with Nash point $(5, 5)$. Threat strategy SQ is anywhere along the line segment AA-AB, with Nash point $(10, 0)$. Rose has a very strong threat in this game.

16.6) a) SQ is $(2.25, 1.80)$; Nash point is $(4.17, 6.27)$.

b) SQ is $(0, 1)$; Nash point is $(3.21, 8.50)$. The security level solution is close to Braithwaite's, but Matthew's strong threat of playing A gives him the advantage in the optimal threat strategy solution.

17.2) 2^{m+l}.

17.3) The Nash solution is at $(1, 1)$, which is most easily implemented as RC. (It could also be implemented as $\frac{1}{3}$PRC + $\frac{2}{3}$PC, and in other ways.)

17.4) The payoff polygon does not pass through the first quadrant, so the solution is at $(0, 0)$. This could be implemented as SQ. (It could also be implemented as PC, and in other ways.)

19.1) a) $(\frac{3}{5})(\frac{3}{7})$AAA + $(\frac{3}{5})(\frac{4}{7})$AAB + $(\frac{2}{5})(\frac{3}{7})$BAA + $(\frac{2}{5})(\frac{4}{7})$BAB $= (2.09, -0.66, -1.43)$

b) Rose A would give Rose $\frac{3}{7}$AAA + $\frac{4}{7}$AAB = 2.14. Rose B would give Rose $\frac{3}{7}$BAA + $\frac{4}{7}$BAB = 2.00. Rose' counter-prudential strategy is A.

c) Colin A; Larry A.

d) No, since each player would be tempted to play counter-prudentially.

e) AAA $= (1, 1, -2)$. Rose and Larry would lose, Colin gain.

19.2) a) A4 is dominated by A1 or A3. No player has a dominant strategy.

b) CBS 1 CBS 2

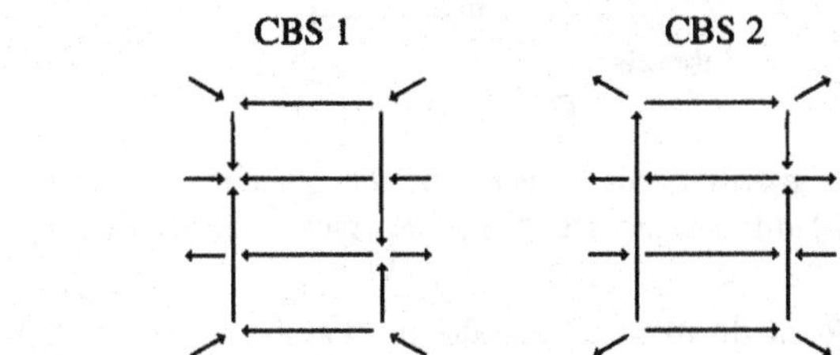

c) Since there is a unique equilibrium at ABC 2–NBC 1–CBS 1, it might be reasonable to predict this as the outcome.

19.3) a) A is a dominant strategy for each player, leading to a unique equilibrium at AAA $= (4, 3, 3)$.

b) The game has a saddle point at ABA $= (1, 2, 7)$.

c) Saddle point at AAA $= (4, 3, 3)$.

d) Saddle point at BAA $= (3, 5, 2)$.

e) Yes, Colin would do worse by joining in a coalition with Larry than by playing alone.

f) Rose prefers Larry, Larry prefers Colin, Colin prefers Rose.

g)
$$v(\phi) = 0 \qquad v(R) = 1 \qquad v(C) = 3 \qquad v(L) = 2$$
$$v(RC) = 8 \qquad v(RL) = 7 \qquad v(CL) = 9 \qquad v(RCL) = 10$$

19.4) a)
$$v(\phi) = v(A) = v(B) = v(C) = v(D) = 0$$
$$v(BC) = v(BD) = v(CD) = 0$$
$$v(AB) = v(AC) = v(AD) = 1$$
$$v(ABC) = v(ABD) = v(ACD) = v(BCD) = v(ABCD) = 1.$$

b)
$$v(\phi) = v(A) = v(B) = v(C) = v(D) = 0$$
$$v(AB) = v(AC) = v(BC) = 200 \qquad v(AD) = v(BD) = v(CD) = 100$$
$$v(ABC) = 600 \qquad v(ABD) = v(ACD) = v(BCD) = 400$$
$$v(ABCD) = 900.$$

19.5)
$$v(\phi) = 0 \qquad v(R) = v(C) = v(L) = 1$$
$$v(RC) = v(RL) = v(CL) = 4 \qquad v(RCL) = 9$$

It does not seem reasonable that in this purely cooperative situation a coalition would play to hold down the payoff of a complementary coalition, since it would hurt itself by doing so. Hence the values for the one and two player coalitions seem too low.

20.1) The Conservatives have a dominant strategy of voting for B. Given that, the game reduces to

Liberals \ Democrats	O	G
G	B ⟶	G
	↓	↑
O	O ⟵	B

One of the two groups needs to persuade the other to vote insincerely. The Democrats tried hard, but unsuccessfully, to persuade Goodell voters to vote for Ottinger.

20.2) In Agenda 2, CR should vote for N, producing N as the winner. In Agenda 3, MR should vote for N and LD should vote for H. Then H would win.

20.3) An agenda producing D as the winner is

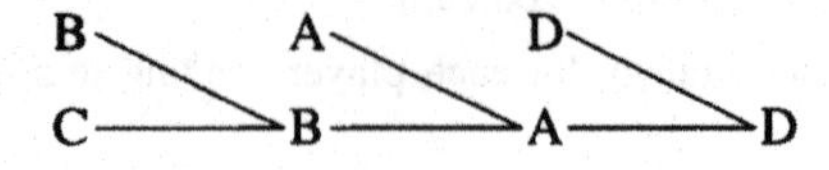

This is unfortunate since the voters *unanimously* prefer C to D, so D is not Pareto optimal.

20.4) MR has straightforward strategy A; MD has straightforward strategy H. Given these, the game reduces to

		LD	
		N	H
CR	A	A ⟶	H
		↑	‖
	N	N ⟵	H

The saddle point produces H as the sophisticated outcome, with the liberal Democrats supporting it.

21.2) If at least three players play C, the C's would wish to stay at C and the D's would wish to switch to C, so the game should move toward an outcome of all C's. If two or fewer players play C, the C's would wish to switch to D and the D's would wish to stay at D, so the game should move toward an outcome of all D's. This kind of game is called a *threshold game*. Think of people deciding whether to attend a party which will be a success only if enough people come.

21.3) a)

		Number of others choosing C					
		0	1	2	3	4	5
Farmer	C	$500	$600	$700	$800	$900	$1000
chooses	D	$800	$1000	$1200	$1400	$1600	$1800
		value of Farmer's cows					

It looks like we would end up with every farmer grazing two $400 cows.

b) Five farmers could gain by restricting themselves to one cow (then worth $900).

c) The one non-cooperative farmer would then get to graze two $900 cows.

22.1)

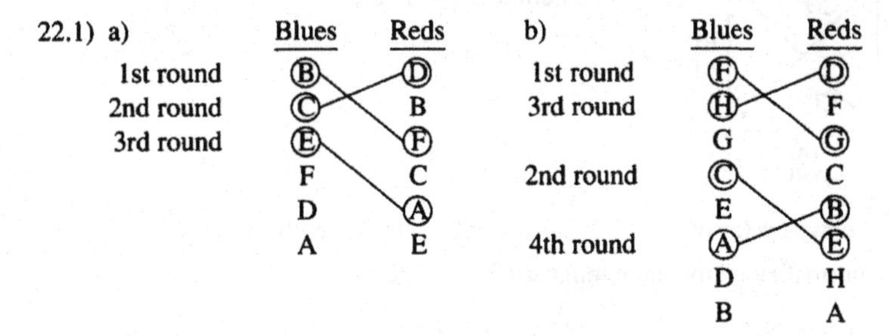

22.4) Reds get E, F; Greens get C, D; Blues get A, B. All teams are better off under optimal play in this ordering than they were under optimal play in the original ordering. In particular, this is true of the Blues. The Blues do better choosing *last* in the draft than choosing first.

22.5) a) One way is to use two copies of the 3-team example, with disjoint player sets. Have each group of 3 teams rank all of its players above all the players of the other group.

b) Here is an example due to Peter Ungar:

	Blues	Reds	Greens	Violets
2nd round	(A)	(E)	C	(H)
	B	F	(F)	G
1st round	(C)	(G)	E	(B)
	D	B	(D)	A
	E	A	A	C
	F	D	B	D
	G	C	G	E
	H	H	H	F

optimal choices

23.2) a) $$v(\phi) = v(1) = v(2) = v(3) = 0 \qquad v(123) = 1$$
$$v(12) = \tfrac{1}{3} \qquad v(13) = \tfrac{1}{2} \qquad v(23) = \tfrac{2}{3}$$

b) The game is inessential. If you try the normalization procedure, every coalition, including N, will get the value 0.

23.3) a) Domination is via $\{23\}$.

b) Domination is via $\{13\}$.

c) **q** dominates **s** via $\{23\}$.

e) $(\frac{11}{32}, \frac{1}{2}, \frac{5}{32})$ dominates **q** via $\{13\}$, or $(\frac{1}{4}, \frac{19}{32}, \frac{5}{32})$ dominates **q** via $\{23\}$.

f) No two person coalition has a value greater than the sum of what its members are getting in **p**.

23.4) 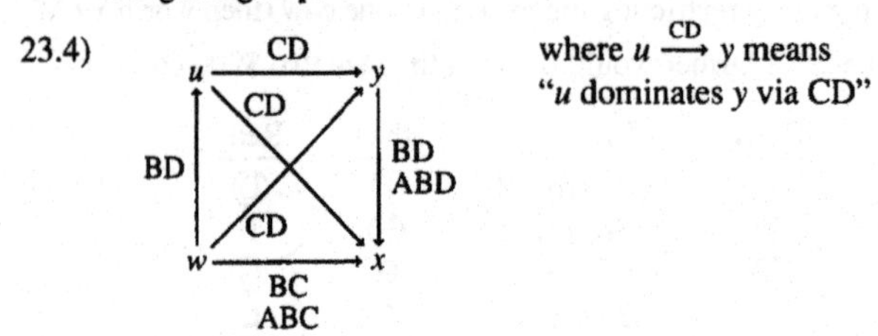

There are two cycles of length three, and one cycle of length four.

23.5) a) "Unnormalizing" the three imputations in $\mathcal{F}$ gives

$$(.515, .915, -.143) \qquad (.515, -4, 3.485) \qquad (-4.4, .915, 3.485)$$

b) Rose and Colin agree to let Larry get 1.03.

23.6) None of the imputations is preferred to any of the others by a coalition which has value greater than zero, so the set is internally stable. Now consider any imputation **y** not in the set. If A gets at least $\frac{2}{3}$, each of the other players must get less than $\frac{1}{3}$, so **s** dominates **y** via BCD. If A gets less than $\frac{2}{3}$, there must be at least one other player who gets less than $\frac{1}{3}$, so one of **p**, **q**, or **r** will dominate **y**. Hence the set is externally stable.

24.1) $$\frac{(e+f+g)-(a+b+c)}{12} < l < \frac{a+b+c}{8} + \frac{e}{2}.$$

24.2) $0 < l < \frac{2n+1}{n^2-1}$.

25.1) The core is a trapezoid with vertices $(3, 2, 1)$, $(3, 3, 0)$, $(2, 4, 0)$, and $(1, 4, 1)$.

25.2) b (BCD gets < 400) and h (ABC gets < 600) are not in the core. The others all are, though not all of them seem very fair.

25.3) As the imputation **x** moves along the curve, the shaded regions sweep out the entire triangle, without overlapping the curve.

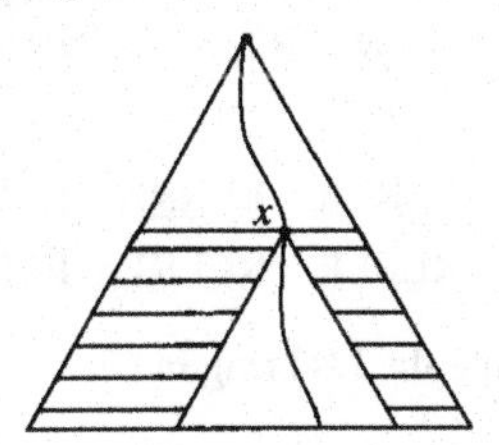

25.4) a) $v(1) = 1$, $v(2) = v(3) = 3$, $v(12) = 5$, $v(13) = v(23) = 6$, $v(123) = 9$ normalizes to $v(1) = v(2) = v(3) = v(23) = 0$, $v(12) = 1$, $v(13) = v(123) = 2$.

b) The core is a line segment between $(1, 0, 1)$ and $(2, 0, 0)$, corresponding to player 1 selling the house to player 3 for a price between \$200,000 and \$300,000.

25.5) a) $(.4, .4, .4, .4, .4)$ is dominated by, for example, $(0, .5, .5, .5, .5)$ via $L_2L_3R_1R_2$.

b) $(.2, .2, .2, .7, .7)$ is dominated by, for example, $(.25, .25, 0, .75, .75)$ via $L_1L_2R_1R_2$.

c) The only undominated imputation is $(0, 0, 0, 1, 1)$, where the right glovers get all the profit. As in the real estate example, the right glovers can play the left glovers against each other.

25.6) a)

Size	If clean up	If pollute
1	−13	−10
2	−22	−20
3	−27	−30
4	−28	−40
5	−25	−50

Coalitions of size 3, 4, and 5 would find it cheaper to clean up than to pollute.

b) $v(1 \text{ player}) = v(2 \text{ players}) = 0$, $v(3 \text{ players}) = 3$, $v(4 \text{ players}) = 12$, $v(5 \text{ players}) = 25$.

c) Check that each 3 player coalition gets at least 3, and each 4 player coalition gets at least 12. $(5, 5, 5, 5, 5)$ corresponds to each village cleaning its own waste. Each village then saves 5 over the cost of cleaning its intake water. $(11, 11, 1, 1, 1)$ corresponds to villages 3, 4, 5 treating their own waste and also paying for the treatment of the waste of villages 1 and 2, and paying those villages a bribe of 1 unit each to allow their wastes to be treated! This hardly seems fair, but it is as cheap for villages 3, 4, and 5 to do that as to pay to clean their intake water if villages 1 and 2 pollute. $(13, 3, 3, 3, 3)$ corresponds to villages 2, 3, 4, 5 treating their own waste, paying for the treatment of village 1's waste, and also paying village 1 a bonus of 3 units to let them do it.

25.8) The core is non-empty if and only if $a + b + c \leqslant 2$.

26.1) a) $\varphi = (\frac{2}{3}, \frac{1}{6}, \frac{1}{6})$, which is not in the core. It is, though, in the symmetric stable set $\mathcal{F} = \{(x_1, x_2, x_3) \mid x_2 = x_3\}$.

b) $(\varphi_W, \varphi_G, \varphi_H) = (.92, 2.02, 2.27)$. The core is empty.

c) $\varphi = (\frac{1}{2}, \frac{1}{6}, \frac{1}{6}, \frac{1}{6})$. The core is empty.

d) $\varphi = (250, 250, 250, 150)$, in the core. The calculation is helped by noting that A, B, C have symmetric roles in the game.

e) $\varphi = (\frac{7}{30}, \frac{7}{30}, \frac{7}{30}, \frac{13}{30}, \frac{13}{30})$, not in the core. An easy way to do this is to write out the 10 orders in which 3 L's and 2 R's can be arranged, and underline the players who bring in $1:

$$\mathrm{LLL\underline{RR}} \quad \mathrm{LL\underline{R}L\underline{R}} \quad \mathrm{LL\underline{RR}L} \quad \mathrm{L\underline{R}LL\underline{R}} \quad \mathrm{L\underline{R}L\underline{R}L}$$

$$\mathrm{L\underline{R}R\underline{L}L} \quad \mathrm{R\underline{L}LL\underline{R}} \quad \mathrm{R\underline{L}L\underline{R}L} \quad \mathrm{R\underline{L}R\underline{L}L} \quad \mathrm{RR\underline{LL}L}$$

The R's bring in $13, while the L's bring in $7.

26.2) $(-1.13, -.72, 1.85)$.

27.1) a) AB, ABC, ABD, ACD, ABCD. Proper, not strong.

b) Same as a). Voting games with the same winning coalitions are *equivalent*.

c) All coalitions containing A and at least two other voters. Proper, not strong (e.g. neither AB nor CDE is winning).

d) AB, BC, CD, and any three or four person coalition. Not proper (both AB and CD are winning), not strong (neither AC nor BD is winning).

27.2) a) and b) $(\frac{7}{12}, \frac{3}{12}, \frac{1}{12}, \frac{1}{12})$

c) $(\frac{6}{10}, \frac{1}{10}, \frac{1}{10}, \frac{1}{10}, \frac{1}{10})$

d) $(\frac{1}{6}, \frac{1}{3}, \frac{1}{3}, \frac{1}{6})$

27.3) N's now have $126/3003 = .042$ of the power.

27.4) President: $50/504 = .099$; Total senators: $204/504 = .405$; Total representatives: $250/504 = .496$.

27.5) a) $\varphi = \frac{1}{18}(4, 4, 4, 1, 1, 1, 1, 1, 1)$.

b) $\varphi = \frac{1}{90}(14, 14, 14, 14, 14, 5, 5, 5, 5)$.

27.6) Bates was right to object. Without Harris, Bates has $\frac{5}{7} = 71\%$ of the power. With Harris, Bates has $\frac{4}{8} = 50\%$ of the power.

27.7) In 1958, $\varphi = (\frac{1}{3}, \frac{1}{3}, \frac{1}{3}, 0, 0, 0)$. In 1964, $\varphi = (\frac{1}{3}, \frac{1}{3}, 0, \frac{1}{3}, 0, 0)$. Banzhaf won his case; the court overturned this scheme and required a power analysis of other weighted voting schemes in New York.

27.8)

		Active Men	
		No bloc	Bloc
Active	No bloc	(.267, .40)	(.16, .63)
Women	Bloc	(.35, .36)	(.273, .55)

Both groups have a dominant strategy of forming a bloc. If they do, they both gain power (the women, not very much).

28.1) a) and b) $(\frac{5}{10}, \frac{3}{10}, \frac{1}{10}, \frac{1}{10})$

c) $(\frac{11}{23}, \frac{3}{23}, \frac{3}{23}, \frac{3}{23}, \frac{3}{23})$

d) $(\frac{1}{6}, \frac{1}{3}, \frac{1}{3}, \frac{1}{6})$

28.2) The P's have 4240 swings; the N's have 840 swings. $\beta_P = 848/5080 = .1669$. $\beta_N = 84/5080 = .0165$.

28.3) Same as the Shapley-Shubik indices.

29.1) $\varphi = (25, 35, 45)$ $\quad \varphi = (35, 45, 55)$.

29.2) A objects against C by proposing $\{AB\}$ and $(11, 49, 0)$.

29.3) For $(15, 65, 55)$, C objects against B by proposing $\{AC\}$ and $(24, 0, 56)$. For $(40, 60, 35)$, C objects against B by proposing $\{AC\}$ and $(44, 0, 36)$.

29.4)

$\{A\}\{B\}\{C\}$	$(0, 0, 0)$
$\{AB\}\{C\}$	$(2, 3, 0)$
$\{AC\}\{B\}$	$(2, 0, 6)$
$\{A\}\{BC\}$	$(0, 3, 6)$
$\{ABC\}$	$(1\frac{2}{3}, 2\frac{2}{3}, 5\frac{2}{3})$

29.5) a) C objects against A by proposing $\{BC\}$ with $(*, 49, 41, *)$. A can try counter-objections with $\{AB\}$, $\{AD\}$ or $\{ABD\}$, but cannot make any of them valid.

b) C counter-objects against A by $\{BC\}$ with $(*, 45, 45, *)$.

c) A counter-objects against C by $\{ABD\}$ with $(30, 45, 0, 45)$.

29.6) For 1, B objects against C by $\{BD\}$ with $(*, 45, *, 45)$. For 7, B objects against D by $\{ABC\}$ with $(35, 50, 35, 0)$. For 8, B objects against C by $\{ABD\}$ with $(35, 50, 0, 35)$.

30.1) b) $\varphi = (.6, .1, .1, .1, .1)$.

30.2) a) H counter-objects with $\{GHJ\}$ and **g**.

b) F counter-objects with $\{FGK\}$ and **d**.

c) F counter-objects with $\{FGK\}$ and **f**.

31.1) a) $\varphi = (35, 45, 55)$ $\quad \nu = (25, 45, 65)$ $\quad \gamma = (28.6, 45, 61.4)$.

b) $\varphi = (1\frac{1}{3}, 2\frac{1}{3}, 3\frac{1}{3})$ $\quad \nu = (\frac{1}{2}, 2\frac{1}{4}, 4\frac{1}{4})$ $\quad \gamma = (\frac{7}{9}, 2\frac{1}{3}, 3\frac{8}{9})$

31.2) a) $\nu = (\frac{1}{3}, \frac{1}{3}, \frac{1}{3}, 0)$.

b) $\nu = (1, 0, 0, 0)$. As a power index, the nucleolus would give all power to voters with vetos, and do other strange things.

31.3) a) $\nu = (1, 0, 0)$ $\quad$ b) $\nu = (0, 0, 0, 1, 1)$
since the nucleolus must be in any non-empty core.

31.4) a) $(1, 2, 3, 4)$ is in the core.

b) $\gamma = (1\frac{3}{7}, 2\frac{1}{7}, 2\frac{6}{7}, 3\frac{4}{7})$. This violates the core constraints for BD and CD.

Index

Look first in Contents